# FROM STATISTICAL PHYSICS TO DATA-DRIVEN MODELLING

# From Statistical Physics to Data-Driven Modelling

## with Applications to Quantitative Biology

Simona Cocco

Rémi Monasson

Francesco Zamponi

# OXFORD
## UNIVERSITY PRESS

Great Clarendon Street, Oxford, OX2 6DP,
United Kingdom

Oxford University Press is a department of the University of Oxford.
It furthers the University's objective of excellence in research, scholarship,
and education by publishing worldwide. Oxford is a registered trade mark of
Oxford University Press in the UK and in certain other countries

Impression: 1

Published in the United States of America by Oxford University Press
198 Madison Avenue, New York, NY 10016, United States of America

British Library Cataloguing in Publication Data
Data available

Library of Congress Control Number: 2022937922

ISBN 978–0–19–886474–5

DOI: 10.1093/oso/9780198864745.001.0001

Printed and bound by
CPI Group (UK) Ltd, Croydon, CR0 4YY

# Contents

# Preface

Today's science is characterised by an ever-increasing amount of data, due to instrumental and experimental progress in monitoring and manipulating complex systems made of many microscopic constituents. While this tendency is true in all fields of science, it is perhaps best illustrated in biology. The activity of neural populations, composed of hundreds to thousands of neurons, can now be recorded in real time and specifically perturbed, offering a unique access to the underlying circuitry and its relationship with functional behaviour and properties. Massive sequencing has permitted us to build databases of coding DNA or protein sequences from a huge variety of organisms, and exploiting these data to extract information about the structure, function, and evolutionary history of proteins is a major challenge. Other examples abound in immunology, ecology, development, etc.

How can we make sense of such data, and use them to enhance our understanding of biological, physical, chemical, and other systems? Mathematicians, statisticians, theoretical physicists, computer scientists, computational biologists, and others have developed sophisticated approaches over recent decades to address this question. The primary objective of this textbook is to introduce these ideas at the crossroad between probability theory, statistics, optimisation, statistical physics, inference, and machine learning. The mathematical details necessary to deeply understand the methods, as well as their conceptual implications, are provided. The second objective of this book is to provide practical applications for these methods, which will allow students to really assimilate the underlying ideas and techniques. The principle is that students are given a data set, asked to write their own code based on the material seen during the theory lectures, and analyse the data. This should correspond to a two- to three-hour tutorial. Most of the applications we propose here are related to biology, as they were part of a course to Master of Science students specialising in biophysics at the Ecole Normale Supérieure. The book's companion website[1] contains all the data sets necessary for the tutorials presented in the book. It should be clear to the reader that the tutorials proposed here are arbitrary and merely reflect the research interests of the authors. Many more illustrations are possible! Indeed, our website presents further applications to "pure" physical problems, e.g. coming from atomic physics or cosmology, based on the same theoretical methods.

Little prerequisite in statistical inference is needed to benefit from this book. We expect the material presented here to be accessible to MSc students not only in physics, but also in applied maths and computational biology. Readers will need basic knowledge in programming (Python or some equivalent language) for the applications, and in mathematics (functional and linear analysis, algebra, probability). One of our major goals is that students will be able to understand the mathematics behind the meth-

---

[1] https://github.com/StatPhys2DataDrivenModel/DDM_Book_Tutorials

ods, and not act as mere consumers of statistical packages. We pursue this objective without emphasis on mathematical rigour, but with a constant effort to develop intuition and show the deep connections with standard statistical physics. While the content of the book can be thought of as a minimal background for scientists in the contemporary data era, it is by no means exhaustive. Our objective will be truly accomplished if readers then actively seek to deepen their experience and knowledge by reading advanced machine learning or statistical inference textbooks.

As mentioned above, a large part of what follows is based on the course we gave at ENS from 2017 to 2021. We are grateful to A. Di Gioacchino, F. Aguirre-Lopez, and all the course students for carefully reading the manuscript and signalling us the typos or errors. We are also deeply indebted to Jean-François Allemand and Maxime Dahan, who first thought that such a course, covering subjects not always part of the standard curriculum in physics, would be useful, and who strongly supported us. We dedicate the present book to the memory of Maxime, who tragically disappeared four years ago.

Paris, January 2022.

Simona Cocco[1], Rémi Monasson[1,2] and Francesco Zamponi[1]
[1] *Ecole Normale Supérieure, Université PSL & CNRS*
[2] *Department of Physics, Ecole Polytechnique*

# 1
# Introduction to Bayesian inference

This first chapter presents basic notions of Bayesian inference, starting with the definitions of elementary objects in probability, and Bayes' rule. We then discuss two historically motivated examples of Bayesian inference, in which a single parameter has to be inferred from data.

## 1.1 Why Bayesian inference?

Most systems in nature are made of small components, interacting in a complex way. Think of sand grains in a dune, of molecules in a chemical reactor, or of neurons in a brain area. Techniques to observe and characterise quantitatively these systems, or at least part of them, are routinely developed by scientists and engineers, and allow one to ask fundamental questions, see figure 1.1:

- What can we say about the future evolution of these systems? About how they will respond to some perturbation, *e.g.* to a change in the environmental conditions? Or about the behaviour of the subparts not accessible to measurements?
- What are the underlying mechanisms explaining the collective properties of these systems? How do the small components interact together? What is the role played by stochasticity in the observed behaviours?

The goal of Bayesian inference is to answer those questions based on observations, which we will refer to as data in the following. In the Bayesian framework, both the

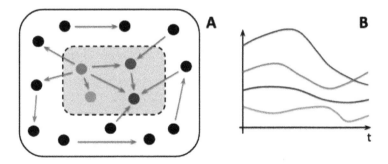

**Fig. 1.1 A**. A large complex system includes many components (black dots) that interact together (arrows). **B**. An observer generally has access to a limited part of the system and can measure the behaviour of the components therein, *e.g.* their characteristic activities over time.

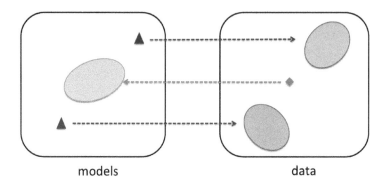

models                                              data

**Fig. 1.2** Probabilistic description of models and data in the Bayesian framework. Each point in the space of model parameters (triangle in the left panel) defines a distribution of possible observations over the data space (shown by the ellipses). In turn, a specific data set (diamond in right panel) corresponding to experimental measurements is compatible with a portion of the model space: it defines a distribution over the models.

data and the system under investigation are considered to be stochastic. To be more precise, we assume that there is a joint distribution of the data configurations and of the defining parameters of the system (such as the sets of microscopic interactions between the components or of external actions due to the environment), see figure 1.2. The observations collected through an experiment can be thought of as a realisation of the distribution of the configurations conditioned by the (unknown) parameters of the system. The latter can thus be inferred through the study of the probability of the parameters conditioned by the data. Bayesian inference offers a framework connecting these two probability distributions, and allowing us *in fine* to characterise the system from the data. We now introduce the definitions and notations needed to properly set this framework.

## 1.2   Notations and definitions

### 1.2.1   Probabilities

We will denote by $y$ a random variable taking values in a finite set of $q$ elements:

$$y \in \{a_1, a_2, \cdots, a_q\} = A \,, \tag{1.1}$$

and by

$$p_i = \text{Prob}(y = a_i) = p(y = a_i) = p(y) \tag{1.2}$$

the probability that $y$ takes a given value $y = a_i$, which will be equivalently written in one of the above forms depending on the context.

For a two dimensional variable

$$\boldsymbol{y} = (y_1, y_2) \in A \times B \tag{1.3}$$

the corresponding probability will be denoted by

$$p_{ij} = \text{Prob}[\boldsymbol{y} = (a_i, b_j)] = p(y_1 = a_i, y_2 = b_j) = p(y_1, y_2) = p(\boldsymbol{y}) . \qquad (1.4)$$

In this case, we can also define the "marginal probabilities"

$$p(y_1) = \sum_{y_2 \in B} p(y_1, y_2) , \qquad p(y_2) = \sum_{y_1 \in A} p(y_1, y_2) . \qquad (1.5)$$

By definition, $y_1$ and $y_2$ are independent variables if their joint probability factorises into the product of their marginals, then

$$p(y_1, y_2) = p(y_1) \times p(y_2) . \qquad (1.6)$$

### 1.2.2    Conditional probabilities and Bayes' rule

The "conditional probability" of $y_1 = a_i$, conditioned to $y_2 = b_j$, is

$$p(y_1|y_2) = \frac{p(y_1, y_2)}{p(y_2)} . \qquad (1.7)$$

Note that this definition makes sense only if $p(y_2) > 0$, but of course if $p(y_2) = 0$, then also $p(y_1, y_2) = 0$. The event $y_2$ never happens, so the conditional probability does not make sense. Furthermore, according to Eq. (1.7), $p(y_1|y_2)$ is correctly normalised:

$$\sum_{y_1 \in A} p(y_1|y_2) = \frac{\sum_{y_1 \in A} p(y_1, y_2)}{p(y_2)} = 1 . \qquad (1.8)$$

Eq. (1.7) allows us to derive Bayes' rule:

$$p(y_1|y_2) = \frac{p(y_1, y_2)}{p(y_2)} = \frac{p(y_2|y_1) \, p(y_1)}{p(y_2)} . \qquad (1.9)$$

This simple identity has a deep meaning for inference, which we will now discuss. Suppose that we have an ensemble of $M$ "data points" $\boldsymbol{y}_i \in \mathbb{R}^L$, which we denote by $Y = \{\boldsymbol{y}_i\}_{i=1,\cdots,M}$, generated from a model with $D$ (unknown to the observer) "parameters", which we denote by $\boldsymbol{\theta} \in \mathbb{R}^D$. We can rewrite Bayes' rule as

$$p(\boldsymbol{\theta}|Y) = \frac{p(Y|\boldsymbol{\theta})p(\boldsymbol{\theta})}{p(Y)} . \qquad (1.10)$$

The objective of Bayesian inference is to obtain information on the parameters $\boldsymbol{\theta}$. We will call $p(\boldsymbol{\theta}|Y)$ the "posterior distribution", which is the object we are interested in: it is the probability of $\boldsymbol{\theta}$ conditioned to the data $Y$ we observe. Bayes' rule expresses the posterior in terms of a "prior distribution" $p(\boldsymbol{\theta})$, which represents our information on $\boldsymbol{\theta}$ prior to any measurement, and of the "likelihood" $p(Y|\boldsymbol{\theta})$ of the model, *i.e.* the probability that the data $Y$ are generated by the model having defining parameters $\boldsymbol{\theta}$. It is important to stress that the likelihood expresses, in practice, a "model of the experiment"; such a model can be known exactly in some cases, but in most situations it is unknown and has to be guessed, as part of the inference procedure. Last of all, $p(Y)$

is called the "evidence", and is expressed in terms of the prior and of the likelihood through

$$p(Y) = \int d\boldsymbol{\theta} \, p(Y|\boldsymbol{\theta}) p(\boldsymbol{\theta}) \ . \tag{1.11}$$

Its primary role is to guarantee the normalisation of the posterior $p(\boldsymbol{\theta}|Y)$.

We will now illustrate how Bayesian inference can be applied in practice with two examples.

## 1.3 The German tank problem

The German tank problem was a very practical issue that arose during World War II. The Allies wanted to estimate the numbers of tanks available to the Germans, in order to estimate the number of units needed for the Normandy landings. Some information was available to the Allies. In fact, during previous battles, some German tanks were destroyed or captured and the Allies then knew their serial numbers, *i.e.* a progressive number assigned to the tanks as they were produced in the factories. The problem can be formalised as follows [1].

### 1.3.1 Bayes' rule

The available data are a sequence of integer numbers

$$1 \le y_1 < y_2 < y_3 < \cdots < y_M \le N \ , \tag{1.12}$$

where $N$ is the (unknown) total number of tanks available to the Germans, and $Y = \{y_1, \cdots, y_M\} \in \mathbb{N}^M$ are the (known) factory numbers of the $M$ destroyed tanks. Note that, obviously, $M \le y_M \le N$. Our goal is to infer $N$ given $Y$.

In order to use Bayes' rule, we first have to make an assumption about the prior knowledge $p(N)$. For simplicity, we will assume $p(N)$ to be uniform in the interval $[1, N_{\max}], p(N) = 1/N_{\max}$. A value of $N_{\max}$ is easily estimated in practical applications, but for convenience we can also take the limit $N_{\max} \to \infty$ assuming $p(N)$ to be a constant for all $N$. Note that in this limit $p(N)$ is not normalisable (it is called "an improper prior"), but as we will see this may not be a serious problem: if $M$ is sufficiently large, the normalisation of the posterior probability is guaranteed by the likelihood, and the limit $N_{\max} \to \infty$ is well defined.

The second step consists in proposing a model of the observations and in computing the associated likelihood. We will make the simplest assumption that the destroyed tanks are randomly and uniformly sampled from the total number of available tanks. Note that this assumption could be incorrect in practical applications, as for example the Germans could have decided to send to the front the oldest tanks first, which would bias the $y_i$ towards smaller values. The choice of the likelihood expresses our modelling of the data generation process. If the true model is unknown, we have to make a guess, which has an impact on our inference result.

There are $\binom{N}{M}$ ways to choose $M$ ordered numbers $y_1 < y_2 < \cdots < y_M$ in $[1, N]$. Assuming that all these choices have equal probability, the likelihood of a set $Y$ given $N$ is

$$p(Y|N) = \frac{1}{\binom{N}{M}} \mathbb{1}(1 \le y_1 < y_2 < \cdots < y_M \le N) , \qquad (1.13)$$

because all the ordered $M$-uples are equally probable. Here, $\mathbb{1}(c)$ denotes the "indicator function" of a condition $c$, which is one if $c$ is satisfied, and zero otherwise.

Given the prior and the likelihood, using Bayes' rule, we have

$$p(N|Y) = \frac{p(Y|N)p(N)}{\sum_{N'} p(Y|N')p(N')} , \qquad (1.14)$$

which for an improper uniform prior reduces to

$$p(N|Y) = \frac{p(Y|N)}{\sum_{N'} p(Y|N')} = \frac{\binom{N}{M}^{-1} \mathbb{1}(N \ge y_M)}{\sum_{N' \ge y_M} \binom{N'}{M}^{-1}} = p(N|y_M; M) . \qquad (1.15)$$

Note that, because $Y$ is now given and $N$ is the variable, the condition $N \ge y_M > y_{M-1} > \cdots > y_1 \ge 1$ is equivalent to $N \ge y_M$, and as a result the posterior probability of $N$ depends on the number of data, $M$ (here considered as a fixed constant), and on the largest observed value, $y_M$, only. It can be shown [1] that the denominator of the posterior is

$$\sum_{N' \ge y_M} \binom{N'}{M}^{-1} = \binom{y_M}{M}^{-1} \frac{y_M}{M-1} , \qquad (1.16)$$

leading to the final expression for the posterior,

$$p(N|y_M; M) = \frac{\binom{N}{M}^{-1} \mathbb{1}(N \ge y_M)}{\binom{y_M}{M}^{-1} \frac{y_M}{M-1}} . \qquad (1.17)$$

Note that for large $N$, we have $p(Y|N) \propto N^{-M}$. Therefore, if $M = 1$, the posterior in Eq. (1.15) is not normalisable if the prior is uniform, and correspondingly Eq. (1.17) is not well defined, see the $M-1$ term in the denominator. Hence, if only one observation is available, some proper prior distribution $p(N)$ is needed for the Bayesian posterior to make sense. Conversely, if $M \ge 2$, the posterior is normalisable thanks to the fast-decaying likelihood, and Eq. (1.17) is well defined, so the use of an improper prior is acceptable when at least two observations are available.

### 1.3.2    Analysis of the posterior

Having computed the posterior probability, we can deduce information on $N$. First of all, we note that $p(N|M, y_M)$ obviously vanishes for $N < y_M$. For large $N$, we have $p(N|M, y_M) \propto N^{-M}$, and recall that we need $M \ge 2$ for the posterior to be well defined. We can then compute:

1. The typical value of $N$, *i.e.* the value $N^*$ that maximises the posterior. As the posterior distribution is a monotonically decreasing function of $N \ge y_M$ we have $N^* = y_M$.

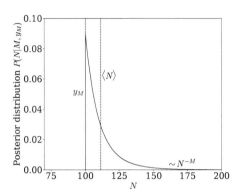

**Fig. 1.3** Illustration of the posterior probability for the German tank problem, for $M = 10$ and $y_M = 100$. The dashed vertical lines locate the typical ($= y_M$) and average values of $N$.

2. The average value of $N$,

$$\langle N \rangle = \sum_N N\, p(N|y_M; M) = \frac{M-1}{M-2}(y_M - 1) \, . \tag{1.18}$$

Note that, for large $M$, we get

$$\langle N \rangle \sim y_M + \frac{y_M - M}{M} + \cdots \, , \tag{1.19}$$

*i.e.* $\langle N \rangle$ is only slightly larger than $y_M$. This formula has a simple interpretation. For large $M$, we can assume that the $y_i$ are equally spaced in the interval $[1, y_M]$. The average spacing is then $\Delta y = (y_M - M)/M$. This formula tells us that $\langle N \rangle = y_M + \Delta y$ is predicted to be equal to the next observation.

3. The variance of $N$,

$$\sigma_N^2 = \langle N^2 \rangle - \langle N \rangle^2 = \frac{M-1}{(M-2)^2(M-3)}(y_M - 1)(y_M - M + 1) \, . \tag{1.20}$$

The variance gives us an estimate of the quality of the prediction for $N$. Note that for large $M$, assuming $y_M \propto M$, we get $\sigma_N/\langle N \rangle \propto 1/M$, hence the relative standard deviation decreases with the number of observations. Notice that Eq. (1.18) for the mean and Eq. (1.20) for the variance make sense only if, respectively, $M > 2$ and $M > 3$. Again, we see a consequence of the use of an improper prior: moments of order up to $K$ are defined only if the number $M$ of observations is larger than $K + 2$.

For example, one can take $M = 10$ observations, and $y_{10} = 100$. A plot of the posterior is given in figure 1.3. Then, $\langle N \rangle = 111.4$, and the standard deviation is $\sigma_N = 13.5$. One can also ask what the probability is that $N$ is larger than some threshold, which is arguably the most important piece of information if you are planning the Normandy landings. With the above choices, one finds

$$p(N > 150|y_{10} = 100) = 2.2 \cdot 10^{-2} \ ,$$
$$p(N > 200|y_{10} = 100) = 1.5 \cdot 10^{-3} \ , \tag{1.21}$$
$$p(N > 250|y_{10} = 100) = 2 \cdot 10^{-4} \ .$$

More generally,

$$p(N > N_{\text{lower bound}}|y_M) \propto \left( \frac{N_{\text{lower bound}}}{\langle N \rangle} \right)^{-(M-1)} . \tag{1.22}$$

## 1.4  Laplace's birth rate problem

We now describe a much older problem, which was studied by Laplace at the end of the eighteenth century. It is today well known that the number of boys and girls is slightly different at birth, although the biological origin of this fact is not fully understood. At his time, Laplace had access to the historical record of the number of boys and girls born in Paris from 1745 to 1770: out of a total number $M = 493472$ of observations, the number of girls was $y = 241945$, while the number of boys was $M - y = 251527$. This observation suggests a slightly lower probability of giving birth to a girl, but could this be explained by random fluctuations due to the limited number of observations?

### 1.4.1  Posterior distribution for the birth rate

Laplace wanted to determine the probability of a newborn baby to be a girl, which is a single parameter, $\theta \in [0, 1]$. To do so, we first need to introduce a prior probability[1] over $\theta$; let us assume $p(\theta) = 1$ to be constant in the interval $\theta \in [0, 1]$ (and zero otherwise), *i.e.* no prior information. To obtain the likelihood, we can assume that each birth is an independent event, in which case the distribution of $y$ conditioned to $\theta$ is a binomial,

$$p(y|\theta; M) = \binom{M}{y} \theta^y (1 - \theta)^{M-y} \ , \tag{1.23}$$

where $M$ is a fixed, known integer. Note that this is a very simple model, because in principle there could be correlations between births (*e.g.* for brothers, twins, etc.). We will discuss later how to assess the quality of a given model. For now we use Bayes' rule to obtain the posterior probability density over the birth rate,

$$p(\theta|y; M) = \frac{p(y|\theta; M)p(\theta)}{p(y; M)} = \frac{\theta^y (1 - \theta)^{M-y}}{\int_0^1 d\theta' (\theta')^y (1 - \theta')^{M-y}} \ . \tag{1.24}$$

Eq. (1.24) is a particular case of the "beta distribution",

$$\text{Beta}(\theta; \alpha, \beta) = \frac{\theta^{\alpha-1}(1 - \theta)^{\beta-1}}{B(\alpha, \beta)} \ ,$$
$$\tag{1.25}$$
$$B(\alpha, \beta) = \int_0^1 d\theta \, \theta^{\alpha-1}(1 - \theta)^{\beta-1} = \frac{\Gamma(\alpha)\Gamma(\beta)}{\Gamma(\alpha + \beta)} \ ,$$

[1]Note that this is now a probability density, because $\theta$ is a continuous variable.

with $\alpha = y + 1$ and $\beta = M - y + 1$, and $\Gamma(x) = \int_0^\infty d\theta\, \theta^{x-1} e^{-\theta}$ stands for the Gamma function. The following properties of the beta distribution are known:

- The typical value of $\theta$, *i.e.* which maximizes Beta, is $\theta^* = \frac{\alpha-1}{\alpha+\beta-2}$.
- The average value is $\langle\theta\rangle = \frac{\alpha}{\alpha+\beta}$.
- The variance is $\sigma_\theta^2 = \frac{\alpha\beta}{(\alpha+\beta)^2(\alpha+\beta+1)}$.

A first simple question is: what would be the distribution of $\theta$ if there were no girls observed? In that case, we would have $y = 0$ and

$$p(\theta|0; M) = (M+1)(1-\theta)^M . \tag{1.26}$$

The most likely value would then be $\theta^* = 0$, but the average value would be $\langle\theta\rangle = 1/M$. This result is interesting: observing no girls among $M$ births should not be necessarily interpreted that their birth rate $\theta$ is really equal to zero, but rather that $\theta$ is likely to be smaller than $1/M$, as the expected number of events to be observed to see one girl is $1/\theta$. From this point of view, it is more reasonable to estimate that $\theta \sim 1/M$, than $\theta = 0$.

In the case of Laplace's data, from the observed numbers, we obtain

$$\theta^* = 0.490291 , \qquad \langle\theta\rangle = 0.490291 , \qquad \sigma_\theta = 0.000712 . \tag{1.27}$$

The possibility that $\theta = 0.5$ seems then excluded, because $\theta^* \sim \langle\theta\rangle$ differs from 0.5 by much more than the standard deviation,

$$|\theta^* - 0.5| \gg \sigma_\theta , \qquad |\langle\theta\rangle - 0.5| \gg \sigma_\theta , \tag{1.28}$$

but we would like to quantify more precisely the probability that, yet, the "true" value of $\theta$ is equal to, or larger than 0.5.

### 1.4.2 Extreme events and Laplace's method

The analysis above suggests that the birth rate of girls is smaller than 0.5. To be more quantitative let us estimate the tiny probability that the observations (number of births) are yet compatible with $\theta > 0.5$. The posterior probability that $\theta > 0.5$ is given by

$$p(\theta > 0.5|y; M) = \int_{0.5}^1 d\theta\, p(\theta|y; M) . \tag{1.29}$$

Unfortunately this integral cannot be computed analytically, but this is precisely why Laplace invented his method for the asymptotic estimation of integrals! Expressing the posterior in terms of $\theta^* = y/M$ instead of $y$, he observed that

$$p(\theta|\theta^*; M) \propto \theta^{M\theta^*}(1-\theta)^{M(1-\theta^*)} = e^{-M f_{\theta^*}(\theta)} , \tag{1.30}$$

where

$$f_{\theta^*}(\theta) = -\theta^* \log\theta - (1-\theta^*)\log(1-\theta) . \tag{1.31}$$

A plot of $f_{\theta^*}(\theta)$ for the value of $\theta^*$ that corresponds to Laplace's observations is given in figure 1.4. $f_{\theta^*}(\theta)$ has a minimum when its argument $\theta$ reaches the typical value $\theta^*$.

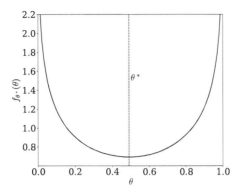

**Fig. 1.4** Illustration of the posterior minus log-probability $f_{\theta^*}(\theta) = -\log p(\theta|\theta^*, M)/M$ for Laplace's birth rate problem.

Because of the factor $M$ in the exponent, for large $M$, a minimum of $f_{\theta^*}(\theta)$ induces a very sharp maximum in $p(\theta|\theta^*; M)$.

We can use this property to compute the normalisation factor at the denominator in Eq. (1.24). Expanding $f_{\theta^*}(\theta)$ in the vicinity of $\theta = \theta^*$, we obtain

$$f_{\theta^*}(\theta) = f_{\theta^*}(\theta^*) + \frac{1}{2}(\theta - \theta^*)^2 f_{\theta^*}''(\theta^*) + O((\theta - \theta^*)^3) , \qquad (1.32)$$

with

$$f_{\theta^*}''(\theta^*) = \frac{1}{\theta^*(1 - \theta^*)} . \qquad (1.33)$$

In other words, next to its peak value the posterior distribution is roughly Gaussian, and this statement is true for all $\theta$ away from $\theta^*$ by deviations of the order of $M^{-1/2}$. This property helps us compute the normalisation integral,

$$\int_0^1 d\theta\, e^{-M\, f_{\theta^*}(\theta)} \sim \int_0^1 d\theta\, e^{-M[f_{\theta^*}(\theta^*) + \frac{1}{2\theta^*(1-\theta^*)}(\theta-\theta^*)^2]}$$
$$\simeq e^{-Mf_{\theta^*}(\theta^*)} \times \sqrt{2\pi\theta^*(1 - \theta^*)/M} , \qquad (1.34)$$

because we can extend the Gaussian integration interval to the whole real line without affecting the dominant order in $M$. We deduce the expression for the normalised posterior density of birth rates,

$$p(\theta|\theta^*; M) \sim \frac{e^{-M[f_{\theta^*}(\theta) - f_{\theta^*}(\theta^*)]}}{\sqrt{2\pi\theta^*(1 - \theta^*)/M}} . \qquad (1.35)$$

In order to compute the integral in Eq. (1.29), we need to study the regime where $\theta - \theta^*$ is of order 1, *i.e.* a "large deviation" of $\theta$. For this, we use Eq. (1.35) and expand it around $\theta = 0.5$, *i.e.*

$$p(\theta > 0.5|\theta^*; M) = \int_{0.5}^{1} d\theta \frac{e^{-M[f_{\theta*}(\theta) - f_{\theta*}(\theta^*)]}}{\sqrt{2\pi\theta^*(1-\theta^*)/M}}$$

$$= \frac{e^{Mf_{\theta*}(\theta^*)}}{\sqrt{2\pi\theta^*(1-\theta^*)/M}} \int_{0.5}^{1} d\theta e^{-M[f_{\theta*}(0.5) + f'_{\theta*}(0.5)(\theta - 0.5) + \cdots]} \quad (1.36)$$

$$\sim \frac{e^{-M[f_{\theta*}(0.5) - f_{\theta*}(\theta^*)]}}{f'_{\theta*}(0.5)\sqrt{2\pi\theta^*(1-\theta^*)M}} .$$

With Laplace's data, this expression for the posterior probability that $\theta \geq 0.5$ could be evaluated to give

$$p(\theta > 0.5|\theta^*; M) \sim 1.15 \cdot 10^{-42} , \quad (1.37)$$

which provides a convincing statistical proof that, indeed, $\theta < 0.5$.

We conclude this discussion with three remarks:

1. The above calculation shows that the posterior probability that $\theta$ deviates from its typical value $\theta^*$ decays exponentially with the number of available observations,

$$p(\theta - \theta^* > a|\theta^*; M) \sim e^{-M[f_{\theta*}(\theta^* + a) - f_{\theta*}(\theta^*)]} . \quad (1.38)$$

   A more general discussion of the large $M$ limit will be given in Chapter 2.

2. The maximum of the function $f$, $f_{\theta*}(\theta^*)$, is equal to the entropy of a binary variable $y = 0, 1$ with corresponding probabilities $1 - \theta^*, \theta^*$. We will introduce the notion of entropy and discuss the general connection with inference in the so-called asymptotic regime (namely, for extremely large $M$) in the next chapter.

3. Note that Eq. (1.36) is strongly reminiscent of a Boltzmann distribution. Assume that a thermal particle moves on a one-dimensional line and feels a potential $U(x)$ depending on its position $x$. Let us call $x^*$ the absolute minimum of $U$. At low temperature $T$, the density of probability that the particle is in $x$ reads (the Boltzmann constant $k_B$ is set to unity here):

$$\rho(x; T) \sim \frac{e^{-(U(x) - U(x^*))/T}}{\sqrt{2\pi \, U''(x^*)/T}} , \quad (1.39)$$

   where the denominator comes from the integration over the harmonic fluctuations of the particle around the bottom of the potential. We see that the above expression is identical to Eq. (1.36) upon the substitutions $\theta^* \to x^*, 0.5 \to x, f_{\theta*} \to U, M \to 1/T$. Not surprisingly, having more observations reduces the uncertainty about $\theta$ and thus the effective temperature.

## 1.5 Tutorial 1: diffusion coefficient from single-particle tracking

Characterising the motion of biomolecules (proteins, RNA, etc.) or complexes (vesicles, etc.) inside the cell is fundamental to the understanding of many biological processes [2]. Optical imaging techniques now allow for the tracking of single particles in real time [3]. The goal of this tutorial is to understand how the diffusion coefficient can be reconstructed from the recordings of the trajectory of a particle, and how the accuracy of the inference is affected by the number of data points (recording length) [4]. The diffusion coefficient depends on the diffusive properties of the environment and on the size of the object. Supposing that the data are obtained in water, from the diffusion coefficient reconstructed from the data the characteristic size of the diffusing object will then be extracted, and a connection with characteristic biological sizes will be made.

### 1.5.1 Problem

We consider a particle undergoing diffusive motion in the plane, with position $r(t) = (x(t), y(t))$ at time $t$. The diffusion coefficient (supposed to be isotropic) is denoted by $\mathcal{D}$, and we assume that the average velocity vanishes. Measurements give access to the positions $(x_i, y_i)$ of the particles at times $t_i$, where $i$ is a positive integer running from 1 to $M$.

**Data:**
Several trajectories of the particle can be downloaded from the book webpage[2], see tutorial 1 repository. Each file contains a three-column array $(t_i, x_i, y_i)$, where $t_i$ is the time, $x_i$ and $y_i$ are the measured coordinates of the particle, and $i$ is the measurement index, running from 1 to $M$. The unit of time is seconds and displacements are in $\mu m$.

**Questions:**

1. Write a script to read the data. Start by the file *dataN1000d2.5.dat*, and plot the trajectories in the $(x, y)$ plane. What are their characteristics? How do they fill the space? Plot the displacement $r_i = \sqrt{x_i^2 + y_i^2}$ as a function of time. Write the random-walk relation between displacement and time in two dimensions, defining the diffusion coefficient $\mathcal{D}$. Give a rough estimate of the diffusion coefficient from the data.

2. Write down the probability density $p(\{x_i, y_i\}|\mathcal{D}; \{t_i\})$ of the time series $\{x_i, y_i\}_{i=1,...,M}$ given $\mathcal{D}$, and considering the measurement times as known, fixed parameters. Deduce, using Bayes' rule, the posterior probability density for the diffusion coefficient, $p(\mathcal{D}|\{x_i, y_i\}; \{t_i\})$.

3. Calculate analytically the most likely value of the diffusion coefficient, its average value, and its variance, assuming a uniform prior on $\mathcal{D}$.

4. Plot the posterior distribution of $\mathcal{D}$ obtained from the data. Compute, for the given datasets, the values of the mean and of the variance of $\mathcal{D}$, and its most likely value. Compare the results obtained with different number $M$ of measures.

[2]https://github.com/StatPhys2DataDrivenModel/DDM_Book_Tutorials

5. Imagine that the data correspond to a spherical object diffusing in water (of viscosity $\eta = 10^{-3}$ Pa s). Use the Einstein-Stokes relation,

$$\mathcal{D} = \frac{k_B T}{6\pi\eta\ell} ,$$ (1.40)

(here $\ell$ is the radius of the spherical object and $\eta$ is the viscosity of the medium) to deduce the size of the object. Biological objects going from molecules to bacteria display diffusive motions, and have characteristic size ranging from nm to $\mu$m. For proteins $\ell \approx 1 - 10$ nm, while for viruses $\ell \approx 20 - 300$ nm and for bacteria, $\ell \approx 2 - 5\,\mu$m. Among the molecules or organisms described in table 1.1, which ones could have a diffusive motion similar to that displayed by the data?

| object | $\ell$ (nm) |
|---|---|
| small protein (lysozime) (100 residues) | 1 |
| large protein (1000 residues) | 10 |
| influenza viruses | 100 |
| small bacteria (e-coli) | 2000 |

**Table 1.1**  Characteristic lengths for several biological objects.

6. In many cases the motion of particles is not confined to a plane. Assuming that $(x_i, y_i)$ are the projections of the three-dimensional position of the particle in the plane perpendicular to the imaging device (microscope), how should the procedure above be modified to infer $\mathcal{D}$?

### 1.5.2  Solution

*Data Analysis.*   The trajectory in the $(x, y)$ plane given in the data file for $M = 1000$ is plotted in figure 1.5**A**. It has the characteristics of a random walk: the space is not regularly filled, but the trajectory densely explores one region before "jumping" to another region.

The displacement $r = \sqrt{x^2 + y^2}$ as a function of time is plotted in figure 1.5**B**. On average, it grows as the square root of the time, but on a single trajectory we observe large fluctuations. The random walk in two dimensions is described by the relation:

$$\langle r^2(t) \rangle = 4\mathcal{D}t ,$$ (1.41)

where $\mathcal{D}$ is the diffusion coefficient whose physical dimensions are $[\mathcal{D}] = l^2 t^{-1}$. Here lengths are measured in $\mu$m and times in seconds. A first estimate of $\mathcal{D}$ from the data can be obtained by just considering the largest time and estimating

$$\mathcal{D}_0 = \frac{r^2(t_{max})}{4t_{max}} ,$$ (1.42)

which gives $\mathcal{D}_0 = 1.20\ \mu\mathrm{m}^2\,\mathrm{s}^{-1}$ for the data set with $M = 1000$. Another estimate of $\mathcal{D}$ can be obtained as the average of the square displacement from one data point

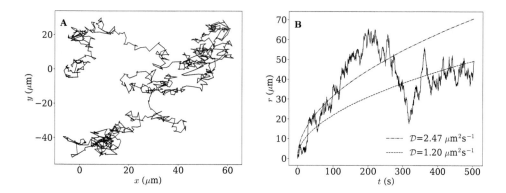

**Fig. 1.5** Data file with $M = 1000$. **A**. Trajectory of the particle. **B**. Displacement from the origin as a function of time.

to the next one divided by the time interval. We define the differences between two successive positions and between two successive recording times

$$\delta x_i = x_{i+1} - x_i , \qquad \delta y_i = y_{i+1} - y_i , \qquad \delta t_i = t_{i+1} - t_i . \qquad (1.43)$$

Note that $i = 1, \ldots, M - 1$. The square displacement in a time step is $\delta r_i^2 = \delta x_i^2 + \delta y_i^2$ and the estimate of $\mathcal{D}$ is

$$\mathcal{D}_1 = \frac{1}{4(M-1)} \sum_{i=1}^{M-1} \frac{\delta r_i^2}{\delta t_i} , \qquad (1.44)$$

giving $\mathcal{D}_1 = 2.47 \ \mu\text{m}^2 \, \text{s}^{-1}$ for the same data set. These estimates are compared with the trajectory in figure 1.5**B**.

*Posterior distribution.* Due to diffusion $\delta x_i$ and $\delta y_i$ are Gaussian random variables with variances $2\mathcal{D} \, \delta t_i$. We have

$$p(\delta x_i | \mathcal{D}; \delta t_i) = \frac{1}{\sqrt{4\pi \mathcal{D} \, \delta t_i}} e^{-\frac{\delta x_i^2}{4 \mathcal{D} \, \delta t_i}} , \qquad (1.45)$$

and $p(\delta y_i | \mathcal{D}; \delta t_i)$ has the same form. The probability of a time series of increments $\{\delta x_i, \delta y_i\}_{i=1,\ldots,M-1}$, given $\mathcal{D}$ is therefore:

$$p(\{\delta x_i, \delta y_i\} | \mathcal{D}; \{\delta t_i\}) = \prod_{i=1}^{M-1} \frac{1}{4\pi \mathcal{D} \, \delta t_i} e^{-\frac{\delta x_i^2}{4 \mathcal{D} \, \delta t_i} - \frac{\delta y_i^2}{4 \mathcal{D} \, \delta t_i}}$$

$$= C e^{-B/\mathcal{D}} \mathcal{D}^{-(M-1)} , \qquad (1.46)$$

where $C = \prod_{i=1}^{M-1} \frac{1}{4\pi \delta t_i}$ and $B = \sum_{i=1}^{M-1} \frac{\delta r_i^2}{4 \, \delta t_i}$. Note that to infer $\mathcal{D}$ we do not need the absolute values of $(x_i, y_i)$, but only their increments on each time interval.

According to Bayes' rule,

$$p(\mathcal{D}|\{\delta x_i, \delta y_i\}; \{\delta t_i\}) = \frac{p(\{\delta x_i, \delta y_i\}|\mathcal{D}; \{\delta t_i\})p(\mathcal{D})}{\int_0^\infty d\mathcal{D}\, p(\{x_i, y_i\}|\mathcal{D}; \{\delta t_i\})p(\mathcal{D})} \ . \tag{1.47}$$

We consider an improper uniform prior $p(\mathcal{D})$ =const. This can be thought as a uniform prior in $[\mathcal{D}_{\min}, \mathcal{D}_{\max}]$, in the limit $\mathcal{D}_{\min} \to 0$ and $\mathcal{D}_{\max} \to \infty$. Thanks to the likelihood, the posterior remains normalisable in this limit.

Note that, introducing

$$\mathcal{D}^* = \frac{B}{M-1} = \frac{1}{4(M-1)} \sum_{i=1}^{M-1} \frac{\delta r_i^2}{\delta t_i} = \mathcal{D}_1 \ , \tag{1.48}$$

we can write the posterior as

$$p(\mathcal{D}|\mathcal{D}^*; M) = \frac{e^{-(M-1)\mathcal{D}^*/\mathcal{D}} \mathcal{D}^{-(M-1)}}{\int_0^\infty d\mathcal{D} e^{-(M-1)\mathcal{D}^*/\mathcal{D}} \mathcal{D}^{-(M-1)}}$$

$$= \frac{e^{-(M-1)\mathcal{D}^*/\mathcal{D}} \mathcal{D}^{-(M-1)} [M-1]\mathcal{D}^*]^{(M-2)}}{(M-3)!} \ , \tag{1.49}$$

where the denominator is easily computed by changing the integration variable to $u = \mathcal{D}^*/\mathcal{D}$. Note that, as in Laplace's problem,

$$p(\mathcal{D}|\mathcal{D}^*; M) \propto e^{-(M-1)f_{\mathcal{D}^*}(\mathcal{D})} \ , \qquad f_{\mathcal{D}^*}(\mathcal{D}) = \frac{\mathcal{D}^*}{\mathcal{D}} + \log \mathcal{D} \ . \tag{1.50}$$

The most likely value of $\mathcal{D}$ is precisely $\mathcal{D}^*$, which is the minimum of $f_{\mathcal{D}^*}(\mathcal{D})$ and the maximum of the posterior, and coincides with the previous estimate $\mathcal{D}_1$.

The average value of $\mathcal{D}$ can also be computed by the same change of variables,

$$\langle \mathcal{D} \rangle = \frac{(M-1)}{(M-3)} \mathcal{D}^* \ , \tag{1.51}$$

and converges to $\mathcal{D}^*$ for $M \to \infty$. The variance of $\mathcal{D}$ is

$$\sigma_{\mathcal{D}}^2 = \frac{(M-1)^2}{(M-3)^2\,(M-4)} (\mathcal{D}^*)^2 \ , \tag{1.52}$$

and it decreases proportionally to $1/M$ for large $M$.

*Numerical analysis of the data.* From the trajectories given in the data files we obtain the results given in table 1.2. An example of the posterior distribution is given in figure 1.6.

Note that for large values of $M$ it is not possible to calculate directly the $(M-3)!$ in the posterior distribution, Eq. (1.49). It is better to use Stirling's formula,

$$(M-3)! \approx \sqrt{2\pi}\, (M-3)^{M-3+1/2}\, e^{-(M-3)} \ , \tag{1.53}$$

| Name-file | $M$ | $\mathcal{D}$ | $\mathcal{D}^*$ | $\langle\mathcal{D}\rangle$ | $\sigma_{\mathcal{D}}$ |
|---|---|---|---|---|---|
| dataN10d2.5.dat | 10 | 2.5 | 1.43 | 1.84 | 0.75 |
| dataN100d2.5.dat | 100 | 2.5 | 2.41 | 2.46 | 0.25 |
| dataN1000d2.5.dat | 1000 | 2.5 | 2.47 | 2.48 | 0.08 |

**Table 1.2** Results for the inferred diffusion constant and its standard deviation (all in $\mu\mathrm{m}^2\,\mathrm{s}^{-1}$), for trajectories of different length $M$.

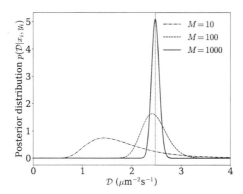

**Fig. 1.6** Posterior distribution for the diffusion coefficient $\mathcal{D}$ given the data, for several values of $M$.

from which one obtains

$$p(\mathcal{D}|B; M) \approx \frac{e^{-B/\mathcal{D}}\,\mathcal{D}^{-(M-1)}\sqrt{M-3}}{\sqrt{2\pi\,e}}\left(\frac{B\,e}{M-3}\right)^{(M-2)}. \tag{1.54}$$

We see that this is a very good approximation for $M = 10$ and we can use it for $M = 100$ and $M = 1000$.

| object | $\ell$ (nm) | $\mathcal{D}$ ($\mu\mathrm{m}^2\,\mathrm{s}^{-1}$) |
|---|---|---|
| small protein (lysozyme) (100 residues) | 1 | 200 |
| large protein (1000 residues) | 10 | 20 |
| influenza viruses | 100 | 2 |
| small bacteria (e-coli) | 2000 | 0.1 |

**Table 1.3** Diffusion constants in water at ambient temperature for several biological objects as in table 1.1, obtained by the Einstein-Stokes relation.

*Diffusion constant and characteristic size of the diffusing object.* The order of magnitude of the diffusion constant can be obtained by the Einstein-Stokes relation: $\mathcal{D} = \frac{k_B T}{6\pi\eta\ell}$, where $\ell$ is the radius of the object (here considered as spherical), and $\eta$ is the viscosity of the medium. Considering the viscosity of the water $\eta = 10^{-3}$ Pa s and $k_B T = 4 \times 10^{-21}$ J, one obtains the orders of magnitude given in table 1.3. Therefore, the data could correspond to an influenza virus diffusing in water.

Ref. [5] reports the following values: for a small protein (lysozyme) $\mathcal{D} = 10^{-6}$ cm$^2$ s$^{-1}$, and for a tobacco virus $\mathcal{D} = 4\ 10^{-8}$ cm$^2$ s$^{-1}$, in agreement with the above orders of magnitude. In Ref. [4], the diffusion coefficient of protein complexes inside bacteria, and with widths approximately equal to $300 - 400$ nm, are estimated to be equal to $\mathcal{D} = 10^{-2}$ $\mu$m$^2$ s$^{-1}$. Differences with the order of magnitude given above are due to the fact that the diffusion is confined and the medium is the interior of the cell, with larger viscosity than water.

# 2

# Asymptotic inference and information

In this chapter we will consider the case of asymptotic inference, in which a large number of data is available and a comparatively small number of parameters have to be inferred. In this regime, there exists a deep connection between inference and information theory, whose description will require us to introduce the notions of entropy, Fisher information, and Shannon information. Last of all, we will see how the maximum entropy principle is, in practice, related to Bayesian inference.

## 2.1 Asymptotic inference

Consider $M$ data points $Y = \{y_i\}_{i=1...M}$, independently drawn from a given likelihood distribution $p(y|\hat{\theta})$. We consider in this chapter the most favourable situation in which $M$ is very large compared to the dimensions $D$ of the parameter vector, $\hat{\theta} \in \mathbb{R}^D$, and $L$ of the data variables $y \in \mathbb{R}^L$. The meaning of "very large" will be made more precise below.

In this section, we begin the discussion by considering the easiest situation, under the following two assumptions:

1. The likelihood $p(y|\theta)$ is exactly known, and the scope of the inference is to infer from the data $Y$ an estimate of the true parameter in the model, $\hat{\theta}$. We will call the estimate $\theta$.
2. The prior $p(\theta)$ over $\theta$ is uniform.

Our aim is to provide some theoretical understanding of the prediction error, which is the difference between $\theta$ and $\hat{\theta}$, and how in the limit of a large number of measurements, $M \to \infty$, this error vanishes asymptotically and is controlled by a simple theoretical bound. We will then discuss non-uniform priors in section 2.1.5, and how to handle cases in which the likelihood is not known in section 2.1.6.

### 2.1.1 Entropy of a distribution

A quantity that plays an extremely important role in asymptotic inference and information theory is the "entropy" of the probability distribution $p$,

$$S(p) = -\sum_i p_i \log p_i = -\sum_{y \in A} p(y) \log p(y) \ . \tag{2.1}$$

This entropy is defined up to a multiplicative constant, which corresponds to the choice of the logarithm base. Common choices are natural or base 2 logs.

For a continuous random variable $y$ taking values in a finite interval $A$, we will denote by $p(y = a)\mathrm{d}a$, or most often simply $p(y)\mathrm{d}y$, the probability that the random variable takes values in $[a, a + \mathrm{d}a]$ or $[y, y + \mathrm{d}y]$, respectively. The quantity $p(y)$ is then called a "probability density". The previous discussion is generalised straightforwardly, and sums over discrete values become integrals over continuous intervals. For example, the entropy is now defined as

$$S(p) = - \int_A \mathrm{d}y \, p(y) \log p(y) . \tag{2.2}$$

## 2.1.2   Cross entropy and Kullback-Leibler divergence

We further need to introduce two important quantities: the cross entropy and the Kullback-Leibler (KL) divergence. Consider two distributions $p(\boldsymbol{y})$ and $q(\boldsymbol{y})$ of a random variable $\boldsymbol{y}$. The cross entropy is defined as

$$S_c(p, q) = - \sum_{\boldsymbol{y}} p(\boldsymbol{y}) \log q(\boldsymbol{y}) = - \langle \log q \rangle_p , \tag{2.3}$$

where the average is over $p$. The name "cross entropy" derives from the fact that this is not properly an entropy, because $p$ and $q$ are different. It coincides with the entropy for $p = q$,

$$S_c(p, p) \equiv S(p) = - \sum_{\boldsymbol{y}} p(\boldsymbol{y}) \log p(\boldsymbol{y}) . \tag{2.4}$$

The KL divergence of $p$ with respect to $q$ is defined as

$$D_{KL}(p\|q) = \sum_{\boldsymbol{y}} p(\boldsymbol{y}) \log \frac{p(\boldsymbol{y})}{q(\boldsymbol{y})} = S_c(p, q) - S(p) . \tag{2.5}$$

An important property of the KL divergence is that it is always positive,

$$D_{KL}(p\|q) \geq 0 , \tag{2.6}$$

where the equality is reached when $p = q$ only. To establish the positivity of $D_{KL}$, consider $z(\boldsymbol{y}) = \frac{q(\boldsymbol{y})}{p(\boldsymbol{y})}$, and note that values of $\boldsymbol{y}$ for which $p(\boldsymbol{y}) = 0$ do not contribute to the sum, because $p(\boldsymbol{y}) \log p(\boldsymbol{y}) \to 0$ when $p(\boldsymbol{y}) \to 0$; hence, we can consider that $p(\boldsymbol{y}) \neq 0$ and $z(\boldsymbol{y})$ is well defined. As $\boldsymbol{y}$ is a random variable, so is $z$, with a distribution induced from the probability $p(\boldsymbol{y})$ over $\boldsymbol{y}$. According to Eq. (2.5), the KL divergence between $p$ and $q$ is $D_{KL}(p\|q) = -\langle \log z(\boldsymbol{y}) \rangle_p$. While the average of $\log z$ is hard to compute, the average value of $z$ is easy to derive:

$$\langle z(\boldsymbol{y}) \rangle_p = \sum_{\boldsymbol{y}} p(\boldsymbol{y}) \, z(\boldsymbol{y}) = \sum_{\boldsymbol{y}} p(\boldsymbol{y}) \frac{q(\boldsymbol{y})}{p(\boldsymbol{y})} = \sum_{\boldsymbol{y}} q(\boldsymbol{y}) = 1 , \tag{2.7}$$

so that

$$-D_{KL}(p\|q) = \langle \log z(\boldsymbol{y}) \rangle_p \leq \langle z(\boldsymbol{y}) \rangle_p - 1 = 0 , \tag{2.8}$$

as a consequence of the concavity inequality $\log x \leq x - 1$. Note that the equality is reached if $z = 1$ for all $\boldsymbol{y}$ such that $p(\boldsymbol{y}) > 0$. As a consequence $D_{KL}(p\|q)$ is positive and vanishes for $p = q$ only.

Hence, the KL divergence gives a measure of the dissimilarity of the distributions $q$ and $p$. Note that this quantity is not symmetric: for generic $p$, $q$, $D_{KL}(p||q) \neq D_{KL}(q||p)$. We will provide a more intuitive interpretation of $D_{KL}$ in section 2.1.4.

### 2.1.3  Posterior distribution for many data

According to Bayes' rule in Eq. (1.10), the posterior distribution $p(\boldsymbol{\theta}|Y)$ is proportional to the likelihood of the model given the data $p(Y|\boldsymbol{\theta})$ (remember the prior over $\boldsymbol{\theta}$ is uniform). If the data are drawn independently, we have

$$p(Y|\boldsymbol{\theta}) = \prod_{i=1}^{M} p(\boldsymbol{y}_i|\boldsymbol{\theta}) = \exp\left( M \times \frac{1}{M} \sum_{i=1}^{M} \log p(\boldsymbol{y}_i|\boldsymbol{\theta}) \right). \tag{2.9}$$

We have rewritten in Eq. (2.9) the product of the likelihoods of the data points as the exponential of the sum of their logarithms. This is useful because, while products of random variables converge badly, sums of many random variables enjoy nice convergence properties. To be more precise, let us fix $\boldsymbol{\theta}$; the log-likelihood of a data point, say, $\boldsymbol{y}_i$, is a random variable with value $\log p(\boldsymbol{y}_i|\boldsymbol{\theta})$, because the $\boldsymbol{y}_i$ are randomly extracted from $p(\boldsymbol{y}_i|\hat{\boldsymbol{\theta}})$. The law of large numbers ensures that, with probability one[1],

$$\frac{1}{M} \sum_{i=1}^{M} \log p(\boldsymbol{y}_i|\boldsymbol{\theta}) \xrightarrow[M\to\infty]{} \int d\boldsymbol{y}\, p(\boldsymbol{y}|\hat{\boldsymbol{\theta}}) \log p(\boldsymbol{y}|\boldsymbol{\theta}) . \tag{2.10}$$

We can then rewrite the posterior distribution for $M \to \infty$ as

$$p(\boldsymbol{\theta}|Y) \propto p(Y|\boldsymbol{\theta}) \approx e^{-M\, S_c(\hat{\boldsymbol{\theta}},\boldsymbol{\theta})} , \tag{2.11}$$

where

$$S_c(\hat{\boldsymbol{\theta}}, \boldsymbol{\theta}) = -\int d\boldsymbol{y}\, p(\boldsymbol{y}|\hat{\boldsymbol{\theta}}) \log p(\boldsymbol{y}|\boldsymbol{\theta}) \tag{2.12}$$

is precisely the cross entropy $S_c(p,q)$ of the true distribution $p(\boldsymbol{y}) = p(\boldsymbol{y}|\hat{\boldsymbol{\theta}})$ and of the inferred distribution $q(\boldsymbol{y}) = p(\boldsymbol{y}|\boldsymbol{\theta})$.

As shown in section 2.1.2, the cross entropy can be expressed as the sum of the entropy and of the KL divergence, see Eq. (2.5)

$$S_c(\hat{\boldsymbol{\theta}}, \boldsymbol{\theta}) = S(\hat{\boldsymbol{\theta}}) + D_{KL}(\hat{\boldsymbol{\theta}}||\boldsymbol{\theta}) , \tag{2.13}$$

where we use the shorthand notation $D_{KL}(\hat{\boldsymbol{\theta}}||\boldsymbol{\theta}) = D_{KL}(p(\boldsymbol{y}|\hat{\boldsymbol{\theta}})||p(\boldsymbol{y}|\boldsymbol{\theta}))$. Due to the positivity of the KL divergence, the cross entropy $S_c(\hat{\boldsymbol{\theta}}, \boldsymbol{\theta})$ enjoys two important properties:

---

[1] If we furthermore assume that the variance of such random variable exists

$$\int d\boldsymbol{y}\, p(\boldsymbol{y}|\hat{\boldsymbol{\theta}}) \left[ \log p(\boldsymbol{y}|\boldsymbol{\theta}) \right]^2 < \infty ,$$

then the distribution of the average of $M$ such random variables becomes Gaussian with mean given by the right hand side of Eq. (2.9) and variance scaling as $1/M$.

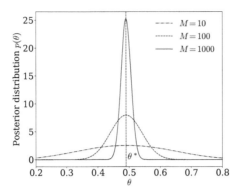

**Fig. 2.1** Illustration of the evolution of the posterior probability with the number $M$ of data, here for Laplace's birth rate problem, see Eq. (1.35). Another example was given in figure 1.6.

- it is bounded from below by the entropy of the ground-truth distribution: $S_c(\hat{\boldsymbol{\theta}}, \boldsymbol{\theta}) \geq S(\hat{\boldsymbol{\theta}})$, and
- it has a minimum in $\boldsymbol{\theta} = \hat{\boldsymbol{\theta}}$.

Therefore, as expected, the posterior distribution in Eq. (2.11) gives a larger weight to the values of $\boldsymbol{\theta}$ that are close to the value $\hat{\boldsymbol{\theta}}$ used to generate the data, for which $S_c(\hat{\boldsymbol{\theta}}, \boldsymbol{\theta})$ reaches its minimum.

### 2.1.4 Convergence of inferred parameters towards their ground-truth values

To obtain the complete expression of the posterior distribution, we introduce the denominator in Eq. (2.11),

$$p(\boldsymbol{\theta}|Y) = \frac{e^{-MS_c(\hat{\boldsymbol{\theta}},\boldsymbol{\theta})}}{\int d\boldsymbol{\theta}\, e^{-MS_c(\hat{\boldsymbol{\theta}},\boldsymbol{\theta})}}\ . \tag{2.14}$$

In the large–$M$ limit the integral in the denominator is dominated by the minimal value of $S_c(\hat{\boldsymbol{\theta}}, \boldsymbol{\theta})$ at $\boldsymbol{\theta} = \hat{\boldsymbol{\theta}}$, equal to the entropy $S(\hat{\boldsymbol{\theta}})$ so we obtain, to exponential order in $M$,

$$p(\boldsymbol{\theta}|Y) \sim e^{-M[S_c(\hat{\boldsymbol{\theta}},\boldsymbol{\theta})-S(\hat{\boldsymbol{\theta}})]} = e^{-MD_{KL}(\hat{\boldsymbol{\theta}}||\boldsymbol{\theta})}\ . \tag{2.15}$$

In figure 2.1 we show a sketch of the posterior distribution for Laplace's problem discussed in section 1.4. The concentration of the posterior with increasing values of $M$ is easily observed.

The KL divergence $D_{KL}(\hat{\boldsymbol{\theta}}||\boldsymbol{\theta}_{hyp})$ controls how the posterior probability of the hypothesis $\boldsymbol{\theta} = \boldsymbol{\theta}_{hyp}$ varies with the number $M$ of accumulated data. More precisely, for $\boldsymbol{\theta}_{hyp} \neq \hat{\boldsymbol{\theta}}$, the posterior probability that $\boldsymbol{\theta} \approx \boldsymbol{\theta}_{hyp}$ is exponentially small in $M$. For any small $\epsilon$,

$$\text{Prob}(|\boldsymbol{\theta} - \boldsymbol{\theta}_{hyp}| < \epsilon) = e^{-MD_{KL}(\hat{\boldsymbol{\theta}}||\boldsymbol{\theta}_{hyp})+O(M\epsilon)}\ , \tag{2.16}$$

and the rate of decay is given by the KL divergence $D_{KL}(\hat{\boldsymbol{\theta}}||\boldsymbol{\theta}_{hyp}) > 0$. Hence, the probability that $|\boldsymbol{\theta}_{hyp} - \boldsymbol{\theta}| < \epsilon$ becomes extremely small for $M \gg 1/D_{KL}(\hat{\boldsymbol{\theta}}||\boldsymbol{\theta}_{hyp})$. The inverse of $D_{KL}$ can therefore be interpreted as the number of data needed to recognise that the hypothesis $\boldsymbol{\theta}_{hyp}$ is wrong.

We have already seen an illustration of this property in the study of Laplace's birth rate problem in section 1.4 (and we will see another one in section 2.1.6). The real value $\hat{\theta}$ of the birth rate of girls was unknown but could be approximated by maximizing the posterior, with the result $\hat{\theta} \approx \theta^* = 0.490291$. We then asked the probability that girls had (at least) the same birth rate as boys, *i.e.* of $\theta_{hyp} \geq 0.5$. The cross entropy of the binary random variable $y_i = 0$ (girl) or 1 (boy) is $S_c(\hat{\theta}, \theta) = f_{\hat{\theta}}(\theta)$ defined in Eq. (1.31) and shown in figure 1.4. The rate of decay of this hypothesis is then

$$D_{KL}(\hat{\theta}||\theta_{hyp}) = f_{\hat{\theta}}(\hat{\theta}) - f_{\hat{\theta}}(\theta_{hyp}) \approx 1.88 \cdot 10^{-4} \,, \tag{2.17}$$

meaning that about 5000 observations are needed to rule out that the probability that a newborn will be a girl is larger than 0.5. This is smaller than the actual number $M$ of available data by two orders of magnitude, which explains the extremely small value of the probability found in Eq. (1.37).

### 2.1.5 Irrelevance of the prior in asymptotic inference

Let us briefly discuss the role of the prior in asymptotic inference. In presence of a non-uniform prior over $\boldsymbol{\theta}$, $p(\boldsymbol{\theta}) \sim \exp(-\pi(\boldsymbol{\theta}))$, Eq. (2.11) is modified into

$$p(\boldsymbol{\theta}|Y) \propto p(Y|\boldsymbol{\theta})p(\boldsymbol{\theta}) \approx \exp\left(-MS_c(\hat{\boldsymbol{\theta}}, \boldsymbol{\theta}) - \pi(\boldsymbol{\theta})\right) \,. \tag{2.18}$$

All the analysis of section 2.1.4 can be repeated, with the replacement

$$S_c(\hat{\boldsymbol{\theta}}, \boldsymbol{\theta}) \to S_c(\hat{\boldsymbol{\theta}}, \boldsymbol{\theta}) + \frac{1}{M}\pi(\boldsymbol{\theta}) \,. \tag{2.19}$$

Provided $\pi(\boldsymbol{\theta})$ is a finite and smooth function of $\boldsymbol{\theta}$, the inclusion of the prior becomes irrelevant for $M \to \infty$, as it modifies the cross entropy by a term of the order of $1/M$. In other words, because the likelihood is exponential in $M$ and the prior does not depend on $M$, the prior is irrelevant in the large $M$ limit. Of course, this general statement holds only if $p(\hat{\boldsymbol{\theta}}) > 0$, *i.e.* if the correct value of $\boldsymbol{\theta}$ is not excluded a *priori*. This observation highlights the importance of avoiding imposing a too restrictive prior.

### 2.1.6 Variational approximation and bounds on free energy

In most situations, the distribution from which data are drawn may be unknown, or intractable. A celebrated example borrowed from statistical mechanics is given by the Ising model over $L$ binary spins (variables) $\boldsymbol{y} = (y_1 = \pm 1, \cdots, y_L = \pm 1)$, whose Gibbs distribution reads (in this section we work with temperature $T = 1$ for simplicity)

$$p_G(\boldsymbol{y}) = \frac{e^{-H(\boldsymbol{y})}}{Z} \,, \quad \text{with} \quad H(\boldsymbol{y}) = -J\sum_{\langle ij \rangle} y_i y_j \,. \tag{2.20}$$

If the sum runs only over pairs $\langle ij \rangle$ of variables that are nearest neighbour on a $d$–dimensional cubic lattice, the model is intractable: the calculation of $Z$ requires

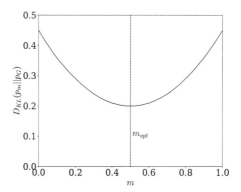

**Fig. 2.2** Illustration of the Kullback-Leibler divergence between the intractable distribution $p_G(\boldsymbol{y})$ and a variational family $p_m(\boldsymbol{y})$.

(for generic Hamiltonians $H$) $2^L$ operations and there is no simple way to bypass this computational bottleneck. Yet, the KL divergence positivity property we have established above can be used to derive bounds on the unknown value of the free energy

$$F = -\log Z \ . \tag{2.21}$$

To do so, let us introduce a family of distributions $p_m(\boldsymbol{y})$ depending on some parameter $m$, which are simple enough to do calculations. The idea is to look for the best approximation to the empirical distribution in this "variational" family, *i.e.* the best value of $m$ (note that $m$ need not be a scalar in the general case). In practice, we may consider the KL divergence between $p_m(\boldsymbol{y})$ and the untractable $p_G(\boldsymbol{y})$,

$$D_{KL}\left(p_m\|p_G\right) = \sum_{\boldsymbol{y}} p_m(\boldsymbol{y}) \log \left[\frac{p_m(\boldsymbol{y})}{p_G(\boldsymbol{y})}\right] \ . \tag{2.22}$$

Using the positivity of the divergence we readily obtain

$$F_m \equiv \langle H(\boldsymbol{y})\rangle_{p_m} - S(p_m) = F + D_{KL}\left(p_m\|p_G\right) \qquad \Rightarrow \qquad F \le F_m \ , \tag{2.23}$$

where $S(p_m)$ is the entropy of $p_m$. Hence, the untractable free energy $F$ is bounded from above by a variational free energy $F_m$ that depends on $m$. Minimising $D_{KL}\left(p_m\|p_G\right)$ (see figure 2.2) or, equivalently, $F_m$, allows us to obtain the best (lowest) upper bound to $F$.

As an illustration let us consider the Ising model in Eq. (2.20) again, defined on a $d$-dimensional cubic lattice. We choose the variational family of factorized distributions,

$$p_m(\boldsymbol{y}) = \prod_{i=1}^{L} \frac{1 + m\, y_i}{2} \ , \tag{2.24}$$

which is much easier to handle because the variables are independent. Note that here $m = \sum_{\boldsymbol{y}} y_i\, p_m(\boldsymbol{y})$ represents the magnetisation of any spin $i$ within distribution $p_m$, and is a real number in the $[-1, 1]$ range. It is straightforward to compute the average value of the Ising energy and of the entropy, with the results

$$\langle H(\boldsymbol{y}) \rangle_{p_m} = -\frac{1}{2} J L \, (2d) \, m^2 \ , \tag{2.25}$$

and

$$S(p_m) = L \left[ - \left( \frac{1+m}{2} \right) \log \left( \frac{1+m}{2} \right) - \left( \frac{1-m}{2} \right) \log \left( \frac{1-m}{2} \right) \right] . \tag{2.26}$$

Minimisation of $F_m$ over $m$ shows that the best magnetisation is solution of the self-consistent equation,

$$m_{opt} = \tanh \left( 2d \, J \, m_{opt} \right) , \tag{2.27}$$

which coincides with the well-known equation for the magnetisation in the mean-field approximation to the Ising model. Hence, the mean-field approximation can be seen as a search for the best independent-spin distribution, in the sense of having the smallest KL divergence with the Ising distribution.

In addition, we know, from section 2.1.4, that the gap $\Delta F = F_m - F$ is related to the quality of the approximation. Suppose somebody gives you $M$ configurations $\boldsymbol{y}$ of $L$ Ising spins drawn from $p_m$, and asks you whether they were drawn from the Gibbs distribution $p_G$ or from $p_m$. Our previous analysis shows that it is possible to answer in a reliable way only if $M \gg 1/\Delta F$. Minimising over $m$ may thus be seen as choosing the value of $m$ that makes the answer as hard as possible, *i.e.* for which the similarity between the configurations produced by the two distributions is the largest.

## 2.2   Notions of information

In this section we introduce several notions of information. We first introduce Fisher information and show that it provides a bound on the quality of a statistical estimator. Then, we introduce other measures of information, such as Shannon information, and we discuss their properties.

### 2.2.1   Statistical estimators

An estimator $\boldsymbol{\theta}^*(Y)$ is a function of the data $Y$ that gives a prediction for the parameters $\boldsymbol{\theta}$ of the model. Examples of estimators based on the posterior distribution are the maximum a posteriori (MAP) estimator, which gives the value of $\boldsymbol{\theta}$ that maximises the posterior,

$$\boldsymbol{\theta}^*_{MAP}(Y) = \text{Argmax}_{\boldsymbol{\theta}} \left[ p(\boldsymbol{\theta}|Y) \right] , \tag{2.28}$$

and the Bayesian estimator, which gives the average of $\boldsymbol{\theta}$ over the posterior,

$$\boldsymbol{\theta}^*_{Bayes}(Y) = \int d\boldsymbol{\theta} \, \boldsymbol{\theta} \, p(\boldsymbol{\theta}|Y) . \tag{2.29}$$

The results of section 2.1 show that when many data are available, $M \to \infty$, these two estimators coincide, and also coincide with the ground truth, $\boldsymbol{\theta}^* = \hat{\boldsymbol{\theta}}$.

More generally, any arbitrary function $\boldsymbol{\theta}^*(Y) : \mathbb{R}^{M \times L} \to \mathbb{R}^D$ can be considered to be an estimator. Even the trivial function $\boldsymbol{\theta}^* \equiv \mathbf{0}$ is an estimator, but it would obviously be a particularly poor estimator if $\hat{\boldsymbol{\theta}} \neq \mathbf{0}$. An important class of estimators is that of unbiased estimators, which are such that:

$$\sum_Y p(Y|\hat{\boldsymbol{\theta}}) \, \boldsymbol{\theta}^*(Y) = \hat{\boldsymbol{\theta}} \ . \tag{2.30}$$

In words, unbiased estimators provide the correct prediction $\hat{\boldsymbol{\theta}}$ if they are fully averaged over the data distribution. By constrast, biased estimators may overestimate or underestimate the true value of the parameters in a systematic way. Note that the MAP and Bayesian estimators defined above are in general biased, except for $M \to \infty$.
   Examples of unbiased estimators are the following.

- Consider $M$ independent and identically distributed random variables $y_i$ with mean $\mu$ and variance $V$. An unbiased estimator for the mean is

$$\mu^*(Y) = \frac{1}{M} \sum_i y_i \ . \tag{2.31}$$

  The average of each $y_i$ is in fact equal to $\mu$, and so is the sample average $\mu^*(Y)$. An unbiased estimator for the variance is $V^*(Y) = \frac{1}{M-1} \sum (y_i - \mu^*(Y))^2$, as can be checked by the reader. Note the presence of the factor $M - 1$ instead of the naive $M$ at the denominator.
- For the German tank problem discussed in section 1.3, the reader can show that the estimator for the number of tanks, $N^* = y_M + \Delta y$ with $\Delta y = y_M - y_{M-1}$ is unbiased.

Unbiased estimators whose variance goes to zero when $M \to \infty$ are particularly desirable. We now discuss a bound on the variance of an unbiased estimator, known as the Cramer-Rao bound.

### 2.2.2   Fisher information and the Cramer-Rao bound

We first consider for simplicity the case $L = 1$, where $\theta$ is a scalar. The posterior $p(\theta|Y) \approx e^{-M D_{KL}(\hat{\theta}||\theta)}$ given in Eq. (2.15) has a sharp peak for $\theta \simeq \hat{\theta}$. We will now investigate the behaviour of $D_{KL}$ close to its minimal value $\hat{\theta}$ where it vanishes. Expanding $D_{KL}$ up to the second order in $\theta$ we obtain

$$D_{KL}(p_{\hat{\theta}}||p_{\theta}) \approx \frac{1}{2}(\theta - \hat{\theta})^2 I_{\boldsymbol{y}}(\hat{\theta}) \ , \tag{2.32}$$

where $I_{\boldsymbol{y}}(\hat{\theta})$ is the Fisher information defined as

$$I_{\boldsymbol{y}}(\hat{\theta}) = -\sum_{\boldsymbol{y}} p(\boldsymbol{y}|\hat{\theta}) \left. \frac{\partial^2}{\partial\theta^2} \log p(\boldsymbol{y}|\theta) \right|_{\theta=\hat{\theta}} \ . \tag{2.33}$$

Asymptotically, the Fisher information gives therefore the variance of $\theta$ when sampled from the posterior distribution of $\theta$,

$$\text{Var}(\theta) = \frac{1}{M\, I_y(\hat{\theta})} \,, \qquad (2.34)$$

so $M$ should be larger than $1/I_y(\hat{\theta})$ for the inference to be in the asymptotic regime.

Outside of the asymptotic regime, the equality in Eq. (2.34) becomes a lower bound for the variance, called the Cramer-Rao bound. For any unbiased estimator $\theta^*(Y)$,

$$\text{Var}(\theta^*) = \sum_Y p(Y|\hat{\theta})[\theta^*(Y) - \hat{\theta}]^2 \geq \frac{1}{I_Y^{total}(\hat{\theta})} \,, \qquad (2.35)$$

where

$$I_Y^{total}(\hat{\theta}) = -\sum_Y p(Y|\hat{\theta}) \left. \frac{\partial^2}{\partial \theta^2} \log p(Y|\theta) \right|_{\theta=\hat{\theta}} \qquad (2.36)$$

is the total Fisher information of the joint distribution of the $M$ data points.

*Proof of the Cramer-Rao bound.* First we note that for an unbiased estimator

$$\sum_Y p(Y|\hat{\theta})(\theta^*(Y) - \hat{\theta}) = 0 \qquad (2.37)$$

is true for each $\hat{\theta}$. We can then differentiate the above equation with respect to $\hat{\theta}$, and get

$$\sum_Y \left( \frac{\partial}{\partial \hat{\theta}} p(Y|\hat{\theta}) \right) (\theta^*(Y) - \hat{\theta}) - 1 = 0 \,, \qquad (2.38)$$

which can be rewritten as

$$\sum_Y p(Y|\hat{\theta}) \left( \frac{\partial}{\partial \hat{\theta}} \log p(Y|\hat{\theta}) \right) (\theta^*(Y) - \hat{\theta}) = 1 \,. \qquad (2.39)$$

Introducing

$$\alpha(Y) = \sqrt{p(Y|\hat{\theta})} \left( \frac{\partial}{\partial \hat{\theta}} \log p(Y|\hat{\theta}) \right) \,,$$
$$\beta(Y) = \sqrt{p(Y|\hat{\theta})} \,(\theta^*(Y) - \hat{\theta}) \,, \qquad (2.40)$$

we can use the Cauchy-Schwarz inequality,

$$\sum_Y |\alpha(Y)|^2 \times \sum_Y |\beta(Y)|^2 \geq |\sum_Y \alpha(Y)\beta(Y)|^2 \,, \qquad (2.41)$$

to obtain

$$\left\{ \sum_Y p(Y|\hat{\theta}) \left( \frac{\partial}{\partial \hat{\theta}} \log p(Y|\hat{\theta}) \right)^2 \right\} \left\{ \sum_Y p(Y|\hat{\theta})(\theta^*(Y) - \hat{\theta})^2 \right\} \geq 1 \,, \qquad (2.42)$$

which gives

$$\text{Var}(\theta^*(Y)) \geq \frac{1}{\sum_Y p(Y|\hat{\theta}) \left( \frac{\partial}{\partial \hat{\theta}} \log p(Y|\hat{\theta}) \right)^2} . \tag{2.43}$$

Observing that

$$I_Y^{total}(\hat{\theta}) = -\sum_Y p(Y|\hat{\theta}) \frac{\partial^2}{\partial \hat{\theta}^2} \log p(Y|\hat{\theta}) = -\sum_Y \frac{\partial^2}{\partial \hat{\theta}^2} p(Y|\hat{\theta}) + \sum_Y \frac{1}{p(Y|\hat{\theta})} \left( \frac{\partial p(Y|\hat{\theta})}{\partial \hat{\theta}} \right)^2$$

$$= \sum_Y p(Y|\hat{\theta}) \left( \frac{\partial \log p(Y|\hat{\theta})}{\partial \hat{\theta}} \right)^2 ,$$

$$\tag{2.44}$$

because the term $\sum_Y \frac{\partial^2}{\partial \hat{\theta}^2} p(Y|\hat{\theta}) = \frac{\partial^2}{\partial \hat{\theta}^2} \sum_Y p(Y|\hat{\theta}) = 0$ vanishes by normalisation of the probability, we obtain the Cramer-Rao bound in Eq. (2.35).

*Fisher information and Cramer-Rao bound in multiple dimensions.* The notion of Fisher information and the Cramer-Rao bound can be extended to a multidimensional parameter $\boldsymbol{\theta} \in \mathbb{R}^L$. The Fisher information is in this case a $L \times L$ matrix:

$$I_{ij} = -\sum_Y p(Y|\hat{\theta}) \frac{\partial^2}{\partial \theta_i \partial \theta_j} \log p(y|\boldsymbol{\theta}) \Big|_{\boldsymbol{\theta}=\hat{\boldsymbol{\theta}}} . \tag{2.45}$$

The fluctuations of the estimators around its unbiased average are quantified by the covariance matrix,

$$C_{ij} = \text{Cov}(\theta_i^*, \theta_j^*) = \sum_Y p(Y|\hat{\theta})(\theta_i^*(Y) - \hat{\theta}_i)(\theta_j^*(Y) - \hat{\theta}_j) . \tag{2.46}$$

The Cramer-Rao bound then states that $M_{ij} = C_{ij} - (I^{-1})_{ij}$ is a positive definite matrix.

### 2.2.3 Properties and interpretation of the Fisher information

We now discuss some additional properties of the Fisher information.

- The Fisher information is additive for independent data: $\log p(Y|\hat{\theta}) = \sum_i \log p(\boldsymbol{y}_i|\theta)$, so that $I_Y^{total} = M I_{\boldsymbol{y}}$, see Eq. (2.36).
- Consider data $Y$ drawn from the distribution $p(Y|\hat{\boldsymbol{\theta}})$. We can define a score for the parameter $\boldsymbol{\theta}$ as

$$s = \frac{\partial}{\partial \boldsymbol{\theta}} \log p(Y|\boldsymbol{\theta}) . \tag{2.47}$$

This quantity tells us how much the data are informative about the value of $\boldsymbol{\theta}$. If the distribution $p(Y|\boldsymbol{\theta})$ does not depend on $\boldsymbol{\theta}$, the score is zero, and no information about the value of $\boldsymbol{\theta}$ can be extracted from the data. The model is in this case non-identifiable: varying the parameters does not affect the distribution of the

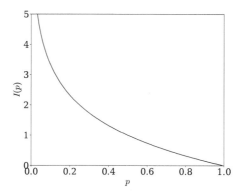

**Fig. 2.3** Illustration of the elementary gain in information following the observation of an event of probability $p$, $I(p) = C \log(1/p)$; here, $C = 1/\log 2$, and the information is expressed in bits.

data. By contrast, if the score is large (positive or negative), then the data bring a lot of information on the parameters. The average value of the score is zero

$$\langle s \rangle = \sum_Y p(Y|\boldsymbol{\theta}) \frac{\partial}{\partial \boldsymbol{\theta}} \log p(Y|\boldsymbol{\theta}) = \sum_Y \frac{\partial p(Y|\boldsymbol{\theta})}{\partial \boldsymbol{\theta}} = 0 \;, \tag{2.48}$$

and the variance of the score is the Fisher information:

$$\langle s^2 \rangle = \sum_Y p(Y|\boldsymbol{\theta}) \left( \frac{\partial}{\partial \boldsymbol{\theta}} \log p(Y|\boldsymbol{\theta}) \right)^2 = I_Y(\boldsymbol{\theta}) \;. \tag{2.49}$$

Hence, larger Fisher information corresponds to larger score variance, and, informally speaking, to better model identifiability.

- The Fisher information is degraded upon transformation: $y \to F(y)$, where $F$ is a deterministic or random function,

$$I_y(\boldsymbol{\theta}) \geq I_F(\boldsymbol{\theta}) \;. \tag{2.50}$$

The extreme case is when $F$ is equal to a constant, in which case all the information is lost. This property is expected to be valid for any measure of information.

### 2.2.4 Another measure of information: Shannon's entropy

The most famous concept of information, besides the Fisher information, was forged by Shannon in 1949. Suppose that you have a series of events $i$ that can happen with probabilities $p_i$, with $i = 1, \cdots, N$. Shannon's idea was to introduce a function $I(p)$ to characterise the information carried by the realisation of one of these events of probability $p$. Informally speaking, observing the realisation of a likely event is not very surprising and thus carries little information. More generally, we may think that the less probable the event is, the larger the information conveyed by its realisation. The function $I(p)$, which characterises the elementary information carried by the observation of an event of probability $p$, should be monotonically decreasing with $p$; such a candidate function is sketched in figure 2.3.

Admissible $I(p)$ must fulfill additional requirements:

- $I(p=1) = 0$: there is no gain of information following the observation of an event happing with certainty.
- $I(p) \geq 0$: the information is always positive.
- $I(p_{12}) = I(p_1) + I(p_2)$ if $p_{12} = p_1 \, p_2$: the information is additive for independent events.
- $I(p_{12}) \leq I(p_1) + I(p_2)$: for correlated events, realisation of the first event is already telling us something about the possible outcomes for the second event. Hence, the information gain from the second event is smaller than what we would have if we had not observed the first event, *i.e.* $I(p_2)$.

These properties, together with a more sophisticated constraint of composability [6], are satisfied by a unique function

$$I(p) = C \log \left( \frac{1}{p} \right) , \qquad (2.51)$$

up to a multiplicative constant that can be chosen as $C = \frac{1}{\log 2}$ to have $I(p)$ measured in bits, *i.e.* $I(\frac{1}{2}) = 1$. The reader can verify that it satisfies all the constraints listed above. In particular, the key property of the logarithmic function in Eq. (2.51) is in that the logarithm of the product of independent variables is equivalent to the sum of the logarithms of the variables, as required by the additivity of the information for independent events.

The Shannon information of a set of events labelled by $i = 1, \cdots, N$ with probabilities $p_i$ is defined as the average gain of information,

$$S(p) = \sum_i p_i \, I(p_i) = \sum_i p_i \, \log_2 \left( \frac{1}{p_i} \right) . \qquad (2.52)$$

Measuring information in bits is convenient when events have binary representations. For instance the Shannon information of a distribution of binary words, *e.g.* $\boldsymbol{y} = 010001\ldots$, of length $L$ and having equal probabilities $p_{unif}(\boldsymbol{y}) = 1/2^L$ is simply $S(p_{unif}) = L$, *i.e.* the length of the binary words. This equality holds for the uniform distribution only; for all other non-uniform distributions, the Shannon information is lower[2].

The Shannon information is nothing but the entropy as defined in statistical physics, with the only difference being the multiplicative prefactor $C$ equal to the

---

[2] To see that, consider the KL divergence between a distribution $p(\boldsymbol{y})$ and the uniform distribution, $p_{unif}(\boldsymbol{y})$:

$$D_{KL}(p\|p_{unif}) = \sum_{\boldsymbol{y}} p(\boldsymbol{y}) \, \log_2[p(\boldsymbol{y})/p_{unif}(\boldsymbol{y})] = L - S(p) . \qquad (2.53)$$

The positivity of the divergence ensures that $S(p) \leq L$.

Boltzmann constant $k_B$ rather than $1/\log 2$. In statistical physics the entropy characterizes the degeneracy of microscopic configurations corresponding to a given macroscopic state. Based on Shannon's interpretation we may see the entropy as how much information we are missing about the microscopic configurations when knowing the macroscopic state of the system only.

### 2.2.5   Mutual information

We now turn to an important and related concept, mutual information. The mutual information of two random variables $x, y$ is defined as the KL divergence between their joint distribution and the product of their marginals:

$$MI(x,y) = D_{KL}(p(x,y)\|p(x)p(y)) = \sum_{x,y} p(x,y) \log\left(\frac{p(x,y)}{p(x)\,p(y)}\right) \qquad (2.54)$$

$$= S[p(x)] + S[p(y)] - S[p(x,y)] \ .$$

Differently from the Kullback-Leibler divergence, the mutual information is a symmetric quantity: $MI(x,y) = MI(y,x)$. In addition it is always positive, and vanishes for independent variables. To better understand its meaning, let us rewrite

$$MI(x,y) = S[p(x)] - S[p(x|y)] = S[p(y)] - S[p(y|x)] \ , \qquad (2.55)$$

where

$$S[p(y|x)] = -\sum_{x,y} p(x,y) \log\left(\frac{p(x,y)}{p(x)}\right) = -\sum_{x} p(x) \sum_{y} p(y|x) \log p(y|x) \qquad (2.56)$$

is called conditional entropy of $y$ at given $x$. The mutual information thus represents the average gain in information over $x$ when $y$ is known, or alternatively, over $y$ when $x$ is known. As we observe the realisation of $x$ our average gain of information is $S[p(x)]$. If we next observe an instantiation of $y$, the mean net gain in information is $S[p(y)] - MI(x,y) \le S[p(y)]$. The sum of these two gains is

$$S[p(x)] + S[p(y)] - MI(x,y) = S[p(x,y)] \ , \qquad (2.57)$$

equal to the average gain in information coming from the joint observation of $x$ and $y$.

## 2.3   Inference and information: the maximum entropy principle

Bayes' rule offers us a powerful framework to infer model parameters from data. However, it presupposes some knowledge over reasonable candidates for the likelihood function, *i.e.* for the distribution $p(Y|\boldsymbol{\theta})$. Is it possible to avoid making such an assumption and also derive the likelihood functional form itself? This is what the maximum entropy principle (MEP), based on the concept of Shannon information, proposes. In few words, MEP consists in inferring maximally agnostic models given what is known about the data, to avoid any superfluous assumption about the data.

As an example, suppose that a scalar variable $y$ takes $N$ possible values with probability $p(y)$. In the absence of any further knowledge about the distribution $p(y)$,

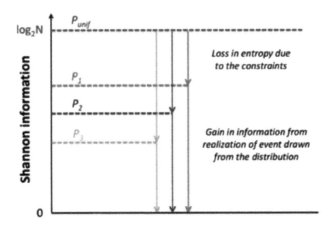

**Fig. 2.4** The maximal entropy model is the distribution $p_1$ with the smallest reduction in Shannon entropy with respect to the uniform model among all possible models $p_1, p_2, p_3, \ldots$ satisfying the constraints imposed by the data.

it is reasonable to suppose that all events are equally likely, *i.e.* $p(y) = \frac{1}{N}$. The entropy of this uniform distribution is $\log_2 N$, and is maximal as we have seen above. This model is, in some sense, maximally "ignorant" and constrains as less as possible the events.

Suppose now that we have some knowledge about the distribution of $y$, for example we are given its average value,

$$m = \langle y \rangle = \sum_y p(y)\, y \;. \tag{2.58}$$

What can we say about $p(y)$? This is in general a very ill-defined question, as there are infinitely numerous distributions $p(y)$ having $m$ as a mean. Yet, not all these distributions $p(y)$ have the same Shannon entropy. The MEP stipulates that we should choose $p(y)$ with maximal entropy among the distributions fulfilling Eq. (2.58). In this way the difference between the entropy of the maximally agnostic (uniform) model and that of the constrained model is minimal, as sketched in figure 2.4. In other words, we have used as little as possible the constraint imposed on us.

In order to find $p(y)$, one then has to maximise its entropy, while enforcing the normalisation constraint together with Eq. (2.58). This can be done by introducing two Lagrange multipliers $\lambda$ and $\mu$. We define

$$S(p, \lambda, \mu) = - \sum_y p(y) \log p(y) - \lambda \left( \sum_y p(y) - 1 \right) - \mu \left( \sum_y p(y)\, y - m \right) \;, \tag{2.59}$$

and write that the functional derivative with respect to the distribution should vanish at the maximum

$$\frac{\delta S(p)}{\delta p(y)} = -\log p(y) - 1 - \lambda - \mu\, y = 0 \; . \tag{2.60}$$

This gives

$$p(y) = \frac{e^{\mu\, y}}{\sum_{y'} e^{\mu\, y'}} \; , \tag{2.61}$$

where we have computed $\lambda$ to fulfill the normalisation constraint. The resulting distribution is then exponential, and the value of $\mu$ is determined by imposing Eq. (2.58).

It is easy to extend the above calculation to higher-dimensional variables $\boldsymbol{y} = (y_1, y_2, ..., y_L)$. For instance, if we are provided with the value of the means

$$m_i = \langle y_i \rangle = \sum_{\boldsymbol{y}} p(\boldsymbol{y})\, y_i \; , \tag{2.62}$$

then the MEP tells us that the most parsimonious use of these constraints corresponds to the distribution

$$p(\boldsymbol{y}) = \frac{e^{\sum_i \mu_i\, y_i}}{\sum_{\boldsymbol{y}'} e^{\sum_i \mu_i\, y_i'}} \; , \tag{2.63}$$

where the $\mu_i$'s should be determined by enforcing the $L$ conditions Eq. (2.62).

The examples above show that application of the MEP under multiple constraints of a fixed observable $\langle \mathcal{O}_k(\boldsymbol{y}) \rangle$, with $k = 1, 2, ..., K$ will lead to an exponential distribution of the form $p(\boldsymbol{y}) \propto \exp[\mu_1 \mathcal{O}_1(\boldsymbol{y}) + \cdots + \mu_K \mathcal{O}_K(\boldsymbol{y})]$. The values of the $\mu_k$'s are determined by imposing that the empirical values $m_k$ of the observables are reproduced by the model[3]. From a practical point of view, this outcome is equivalent to a Bayesian inference approach, in which an exponential model is assumed, and the parameters $\mu_k$ are determined through likelihood maximisation.

The MEP has a long and deep story in statistical mechanics. In 1957 Jaynes [7] showed that it could be used to derive all thermodynamical ensembles. For instance, the Gibbs distribution of the canonical ensemble is determined by a MEP in which the observable is the average energy of configurations, $\mathcal{O}(\boldsymbol{y}) = E(\boldsymbol{y})$, and the associated Lagrange parameter $\mu$ is related to the inverse temperature $T$ through $\mu = -1/(k_B T)$. Similarly, the grand-canonical ensemble is obtained when applying the MEP under the constraints that both the average energy and the average number of particles are fixed. The Lagrange parameter enforcing the latter constraint is related to the chemical potential.

---

[3] As an illustration, assume we are given the first and second moments, respectively, $m_1$ and $m_2$ of a random scalar variable $y$. According to the MEP the log-probability of this variable is a linear combination of $\mathcal{O}_1(y) = y$ and $\mathcal{O}_2(y) = y^2$, *i.e.* the maximal entropy distribution is Gaussian.

## 2.4   Tutorial 2: entropy and information in neural spike trains

Understanding the neural code and information transmission is of fundamental importance in neuroscience. To such aim, neural recordings have been performed on neurons that are sensitive to a given stimulus (*e.g.* neurons in the visual system that are sensitive to visual stimuli, such as retina ganglion cells or V1 neurons), and the variability and reproducibility of neural responses have been quantified with ideas from information theory [8–10]. The goal of this tutorial is to understand how the variability and reproducibility of responses of retinal ganglion cells to natural visual stimuli can be characterised in terms of information that a spike train provides about a stimulus.

### 2.4.1   Problem

The spiking activity of a population of $L = 40$ retina ganglion cells in response to visual stimuli was recorded [11]. During the recording a natural movie, lasting $T = 26.5$ s, is presented $N_r = 120$ times. We want to analyse the information and noise content of the spike train for each neuron, as a function of its spiking frequency.

**Data:**
The spiking times can be downloaded from the book webpage[4], see tutorial 2 repository. Data are taken from Ref. [11]. The data file contains a one-column array $(t_i)$, where $t_i$ is the spiking time of neuron $i$ in seconds and ranging between 0 and $N_r \times T = 3180$ s. They are separated by the number 4000 followed by the neuron label going from 4 to 63 (not to be confused with the neuron index $i$, running from 1 to $L = 40$).

**Questions:**
In the tutorial 2 repository, you are given a Jupyter notebook *tut2_start.ipynb* that reads the data and makes a raster plot of the spikes of the first neuron, for the first 10 seconds and the first 10 repetitions of the stimulus. Complete the notebook or write your own code to answer the following questions.

1. **Entropy of a Poisson process.** To analyse the spike train of a single neuron, we discretise the time in elementary time-bin windows of length $\Delta t$, where typically $\Delta t = 10$ ms, and translate the activity of the neuron in a time bin into a binary symbol $\sigma \in \{0, 1\}$. Here, $\sigma = 0$ corresponds to no spike, while $\sigma = 1$ corresponds to *at least* one spike.

   1a. Theory: Consider the activity of the neuron as a Poisson process with frequency $f$: write the entropy $S_p(f, \Delta t)$ and the entropy rate (entropy per unit of time) of the Poisson process as a function of the frequency $f$ and of the time-bin width $\Delta t$.

   1b. Numerics: Plot the entropy $S_p(f, \Delta t)$ and the entropy rate obtained in point 1a in the range of experimentally observed frequencies, *e.g.* $f = 4$ Hz, as a function of $\Delta t$.

2. **Information conveyed by the spike train about the stimulus.** To study the reproducibility of the neural activity following a stimulus, we consider words

---

[4]https://github.com/StatPhys2DataDrivenModel/DDM_Book_Tutorials

extending over $\ell$ consecutive time bins, *i.e.* of $\ell$ symbols. For $\ell = 100$ ms and $\Delta t = 10$ ms, there are $N_W = 2^{10} = 1024$ possible such words representing the activity of a neuron over $\ell \times \Delta t = 100$ ms. As explained in Ref. [8] one can extract from the data the probability of each possible word $W$ and then estimate the entropy of this distribution:

$$S_{total} = -\sum_W p(W) \, \log_2 p(W) \text{ bits .} \tag{2.64}$$

Moreover, to assess the reproducibility of the response in a recording in which the same stimulus is presented many times, the probability of the occurrence $p(W|b)$ of a word $W$ starting in a particular time bin $b$ relative to the stimulus can be estimated. These distributions (one for each time bin index $b$) have entropies

$$S(b) = -\sum_W p(W|b) \, \log_2 p(W|b) \text{ bits .} \tag{2.65}$$

The average of these entropies defines the so-called noise entropy,

$$S_{noise} = \langle S(b) \rangle_b \text{ ,} \tag{2.66}$$

where $\langle \ldots \rangle$ denotes the average over all possible times with time resolution $\Delta t$. $S_{noise}$ is called noise entropy because it reflects the non-deterministic response to the stimulus. If the noise entropy is zero, the response is deterministic: the same word will be repeated after each given stimulus. The average information $I$ that the spike train provides about the stimulus is defined as $I = S_{total} - S_{noise}$.

2a. Theory: The probability $p(W|b)$ can be thought of as the probability of the random variable $W$ conditioned to the stimulus at time $b \times \Delta t$, itself considered as a random variable; different times $t$ correspond to different random draws of the stimulus. Under this interpretation, show that $I$ is the mutual information between $W$ and the stimulus, and as a consequence $I \geq 0$.

2b. Numerics: What is the distribution of $S_{total}$, $S_{noise}$ and $I$ over the 40 neural cells and $\ell = 10$ time bins? What is, on average, the variability of the spike trains used to convey information about the stimulus? Plot $S_{total}(i)$, $S_{noise}(i)$, $I(i)$, and the ratio $I(i)/S_{total}(i)$ for the $i = 1 \ldots 40$ neurons as a function of their spiking frequencies $f_i$. Compare $S_{total}(i)$ with the entropy of a Poisson process.

3. **Error on the estimate of $I$ due to the limited number of data.** How does $I$ depend on the number of repetitions of the experiment used in the data analysis and on $\ell$? Check if the quantity of information per second converges. Are 120 repetitions and $\ell = 10$ enough?

4. **Time-dependence of the neuronal activity.** Plot the time-bin dependent entropy $S(b)$ for one neuron. For the same neuron, compute the average spiking frequency $f(b)$ in a small time interval around time bin $b$. Compare the frequency modulation $f(b)$ with $S(b)$. Discuss the relation between $S(b)$ and the entropy of a Poisson process with modulated frequency $f(b)$.

### 2.4.2 Solution

**Analytical calculations:**

1a. Consider a Poisson process with frequency $f$ in a time bin $\Delta t$. The probability that the Poisson process emits no spike in the time interval $\Delta t$, that is of, $\sigma = 0$, is

$$p(\sigma = 0) = e^{-f\,\Delta t}, \quad \text{hence} \quad p(\sigma = 1) = 1 - e^{-f\,\Delta t}. \quad (2.67)$$

The entropy in bits of such binary variable is

$$S_p(f, \Delta t) = -\sum_{\sigma} p(\sigma) \log_2 p(\sigma)$$
$$= \frac{e^{-f\Delta t}\, f\,\Delta t - (1 - e^{-f\Delta t})\, \log(1 - e^{-f\,\Delta t})}{\log 2}. \quad (2.68)$$

The entropy per unit of time is simply $S_p(f, \Delta t)/\Delta t$.

2a. We consider a discrete time index of bins, $b = 1, 2, \cdots, B = T/\Delta t$. Now, consider that a time bin $b$ (and hence a corresponding stimulus) is chosen randomly with uniform probability $p(b) = 1/B$. We call $p(b, W)$ the joint probability of choosing a bin $b$ and a word $W$, and then $p(W|b) = p(W, b)/p(b)$. We have

$$p(W) = \sum_b p(W|b)p(b), \quad (2.69)$$

and

$$S_{noise} = \sum_b p(b)S(b), \quad S(b) = -\sum_W p(W|b) \log_2 p(W|b). \quad (2.70)$$

We can therefore identify the noise entropy with the conditional entropy, $S_{noise} = S[p(W|b)]$, as defined in Eq. (2.56). If the relation between $b$ and $W$ is deterministic, then $S_{noise} = 0$. Therefore

$$S_{total} - S_{noise} = -\sum_b p(b) \sum_W \left[ p(W|b)\, \log_2 p(W) - p(W|b)\, \log_2 \frac{p(W, b)}{p(b)} \right]$$
$$= \sum_{W,b} p(W, b) \log_2 \left[ \frac{p(W, b)}{p(W)\, p(b)} \right] = I(W, b). \quad (2.71)$$

The average information $I$ is the mutual information between the stimulus and the neural activity, as defined in Eq. (2.54).

**Data analysis:**

The Jupyter notebook *tutorial2.ipynb* contains as parameters the time bin $\Delta t$, the length of the words $\ell_s$, and the number of repetitions $N_{rp}$ (out of the total $N_r = 120$) that are considered to calculate the entropy.

The raster plot of data for the first neuron is shown in figure 2.5 for the first 10 seconds of recording and the first 10 repetitions of the stimulus. It can be seen that the neuron fires in a quite precise and reproducible way during the stimulations.

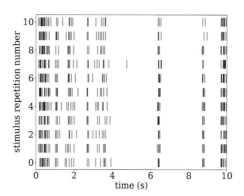

**Fig. 2.5** Raster plot of the activity of the first recorded neuron, in the first 10 seconds of the recording, for the first 10 repetitions of the stimulus.

1b. The entropy $S_p$ and the entropy rate $S_p/\Delta t$ are plotted in figure 2.6 for a cell spiking at $f = 4$ Hz as a function of $\Delta t$. The entropies have a non-trivial behaviour upon changing $\Delta t$. The maximum of the entropy is reached for $\Delta t$ such that $p(\sigma = 0) = p(\sigma = 1) = 0.5$; for other values of $\Delta t$ the entropy decreases. The entropy per unit of time decreases upon increasing $\Delta t$.

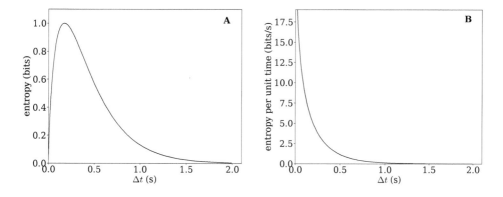

**Fig. 2.6** Entropy (**A**) and entropy per unit time (**B**) of a Poissionian spiking event in a time bin $\Delta t$ for a frequency $f = 4$ Hz as a function of the time bin $\Delta t$

2b. $S_{total}(i)$, $S_{noise}(i)$ and $I(i)$, for the recorded neurons $i = 1, \cdots, 40$ as a function of their frequencies are plotted in figure 2.7, for $\Delta t = 10$ ms and words of duration $\ell_s = 100$ ms, corresponding to $\ell = 10$ binary symbols.
As shown in figure 2.7**A**, the total entropy rate varies between 1 and 25 bits/s. Note that if the neurons were following a Poisson process at frequency $f_i$, then each word would be composed by independent bits, and the total entropy of

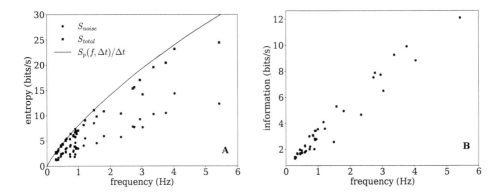

**Fig. 2.7** Total entropy and noise entropy (**A**) and information rate (**B**) as a function of the cell frequency for $\Delta t = 10$ ms, $\ell = 10$. In panel **A**, the full line is the entropy per unit time of a Poisson process, $S_p(f, \Delta t = 10\,\mathrm{ms})$.

neuron $i$ would be

$$S_{total} = \ell\, S_p(f_i, \Delta t) \qquad \Rightarrow \qquad \frac{S_{total}}{\ell \times \Delta t} = \frac{S_p(f_i, \Delta t)}{\Delta t}\,. \tag{2.72}$$

This curve is reported as a full line in figure 2.7**A**, and provides an upper bound to the total entropy. Indeed, correlations between spikes lower the entropy with respect to a purely Poisson process.

The noise entropy is approximatively half of the total entropy, and therefore the information rate $I = (S_{total} - S_{noise})/\Delta t$ varies between 2 and 12 bits/s, see figure 2.7**B**, and is roughly proportional to the neuron spiking frequency. The more active cells have larger entropy and information rate. The order of magnitude of the entropies and information rates are in agreement with Ref. [9, figure 3] which also studied recordings in the retina, for ganglion cells spiking at the same frequencies.

Also, in good agreement with Ref. [8], in which a motion-sensitive neuron in the fly visual system is studied, the noise entropy is of the order of half the total entropy. So, on average, half of the variability of the spike train is used to convey information about the stimulus. The ratio $I/S_{total}$ is shown in figure 2.8.

3. One can vary the number of repetitions of the experiment used in the data analysis: $N_r = 1$ can be used to check that the noise entropy is zero, while the total entropy is not. Moreover one can study the dependence of the results with the number of repetitions and check if $N_r = 120$ is enough to have a stable result. In figure 2.9**A**, the information $I$ is plotted versus $N_r$ for one neuron (the first recorded).

The entropy should reach a constant value in the infinite word-length limit, *i.e.* for the value of $\ell_s = \ell \times \Delta t$ at which the process is uncorrelated. As an example we show in figure 2.9**B** the evolution of $I$ with $\ell_s$, for the first recorded neuron. Very similar results are obtained, in both cases, for the other neurons.

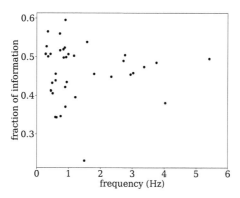

**Fig. 2.8** Ratio of $I/S_{tot}$ as a function of the cell frequency for $\Delta t = 10$ ms, $\ell = 10$.

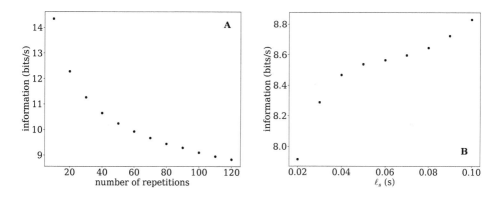

**Fig. 2.9** Dependence of $I$ on the parameters for the first recorded neuron. **A.** $I$ vs. number of repetitions of the experiment. **B.** $I$ vs. duration of the words $\ell_s = \ell \times \Delta t$.

4. In figure 2.10 the time-dependent noise entropy $S(t)$ (not averaged over $t$) for the first neuron is plotted versus time. Also, the time-dependent frequency (averaged over time intervals $[t, t+T]$) is plotted versus $t$. If the neuron were Poissonian, then (similarly to what was discussed above) one would have $S(t) = \ell S_p[f(t), \Delta t]$. This quantity is also reported in the figure. The three quantities are strongly correlated. For better visualisation, in figure 2.11 we show parametric plots, over time, of $S(t)$ versus $f(t)$ and of $S(t)$ versus $\ell S_p[f(t), \Delta t]$. We can conclude that even if most of the variation of $S(t)$ is directly related to the frequency modulation $f(t)$, we have $S(t) < \ell S_p[f(t), \Delta t]$ which implies that some additional information is contained in the spike trains.

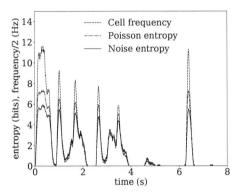

**Fig. 2.10** Time-bin dependent entropy $S(b)$ versus time $t = b \times \Delta t$ for the first recorded neuron, together with the frequency $f(b)$ of the cell (averaged over intervals $[t, t + T]$) and the corresponding Poisson entropy $\ell \, S_p[f(b), \Delta t]$. Here $\Delta t = 10$ ms and $T = 100$ ms.

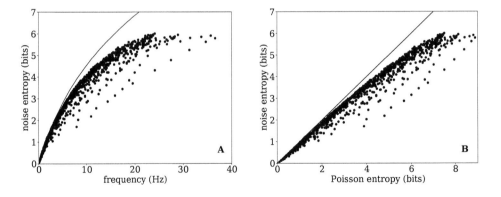

**Fig. 2.11** Parametric plots over time of the time-dependent properties of the first recorded neuron, here for $\Delta t = 10$ ms and $T = 100$ ms. **A.** $S_{noise}$ versus $f$; the full line is $\ell S_p(f, \Delta t)$. **B.** $S_{noise}$ versus $\ell \, S_p(f, \Delta t)$; the full line is $y = x$.

# 3

# High-dimensional inference: searching for principal components

Figure 3.1 graphically summarises what we have done in the first two chapters in terms of complexity of models to be inferred, loosely expressed here as the number of defining parameters, and of the size of the data set. While we have focused so far on how to infer simple models (with one or few parameters) from few or many data, we are going to address now a more complex situation, where the dimensionality of the unknown parameter vector $\boldsymbol{\theta}$ is comparable to the number of available data. This situation, called high-dimensional inference, is relevant in many practical applications of statistical inference methods. We will concentrate on one of them, called principal component analysis (PCA), in this chapter. Other applications will be developed throughout the book.

## 3.1 Dimensional reduction and principal component analysis

Many data are intrinsically high dimensional. For instance, genomic sequences can include millions or billions of nucleotides, each taking one out of four values. Images are defined by the colour code of millions of pixels. Each data point is then represented by a vector in a very high-dimensional space, and capturing the statistical relations or similarities between these points is very hard. A natural idea, called dimensional reduction, is to look for a mapping $\boldsymbol{y} \to \boldsymbol{y}'$, in which the dimension of data is drastically reduced from $L \gg 1$ to $L' \sim 1$. In practice $L'$ is chosen to be 1, 2 or 3, so that the mapped data points can be easily visualised. The mapping must be done in a meaningful way, *i.e.* to lose as little as possible of the data structure. Principal component analysis (PCA) is one way to do this mapping, as we will explain below. Before doing so, we introduce a mathematical framework that will allow us to interpret PCA in a Bayesian inference setting.

### 3.1.1 Multivariate Gaussian distributions

Consider a multivariate Gaussian distribution over some zero mean variables $\boldsymbol{y} = \{y_i\}$, with $(i = 1, \ldots, L)$,

$$p(\boldsymbol{y}|\widehat{T}) = \frac{\sqrt{\det \widehat{T}}}{(2\pi)^{L/2}} \, e^{-\frac{1}{2}\boldsymbol{y}^T \cdot \widehat{T} \cdot \boldsymbol{y}} \,, \quad \text{where} \quad \boldsymbol{y}^T \cdot \widehat{T} \cdot \boldsymbol{y} = \sum_{ij} y_i \, \widehat{T}_{ij} \, y_j \,. \tag{3.1}$$

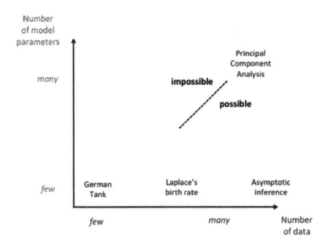

**Fig. 3.1** Trade-off between model complexity and data availability. The German tank problem of Chapter 1 consisted in inferring one parameter from a few observations (bottom left corner). A very favourable situation is when the number of data gets large, while the dimensionality of $\theta$ remains small, as in Laplace's birth rate problem and in the asymptotic inference framework of Chapter 2. In this chapter we will focus on the issue of extracting principal components from many high-dimensional data, and determine the regions in the plane where inference is possible and where it is not.

In the definition above, $\widehat{T}$ is a $L \times L$ positive definite matrix, called precision matrix; it coincides with minus the coupling matrix used in statistical mechanics. Its coefficients, $\widehat{T}_{ij}$, define the graph of dependences between the variables, see figure 3.2 for an example in dimension $L = 3$. In particular, the conditional probability of $y_i$ depends only of the other variables $y_j$'s such that $T_{ij} \neq 0$. Diagonal matrices $\widehat{T}$ define empty graphs, and therefore correspond to independent variables. The corresponding density of probability becomes factorized,

$$p(\boldsymbol{y}|\widehat{T} = \text{diagonal}(\widehat{T}_{11}, \widehat{T}_{22} \ldots, \widehat{T}_{LL})) = \prod_{i=1}^{L} \sqrt{\frac{\widehat{T}_{ii}}{2\pi}} \exp\left(-\frac{\widehat{T}_{ii}}{2} y_i^2\right). \qquad (3.2)$$

In this case, the inverse of $\widehat{T}_{ii}$ is simply the variance of $y_i$. For generic, non-diagonal precision matrices, the inverse of the precision matrix is the covariance matrix of the variables:

$$\widehat{C}_{ij} \equiv \int d\boldsymbol{y}\, P(\boldsymbol{y}|\widehat{T})\, y_i y_j = [\widehat{T}^{-1}]_{ij}. \qquad (3.3)$$

In particular, the variance of $y_i$ is the $i^{th}$ element on the diagonal of $\widehat{T}^{-1}$.

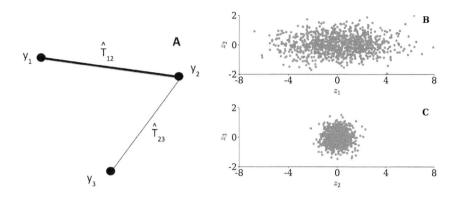

**Fig. 3.2 A**. Graph of dependences for the $3 \times 3$ precision matrix $\widehat{T}$ with diagonal entries $\widehat{T}_{11} = 3.5, \widehat{T}_{22} = 1.5, \widehat{T}_{33} = 2.5$ and off-diagonal entries $\widehat{T}_{12} = -2, \widehat{T}_{23} = 0.5, \widehat{T}_{13} = 0$ (the width of the links is proportional to the absolute values of the corresponding coefficients). Notice that the conditional probability of $y_1$ given $y_2, y_3$ is in fact independent of $y_3$ since $\widehat{T}_{13} = 0$. **B, C**. The eigenvalues of the covariance matrix $\widehat{C} = \widehat{T}^{-1}$ are $\widehat{\lambda}_1 \simeq 5.40, \widehat{\lambda}_2 \simeq 0.39, \widehat{\lambda}_3 \simeq 0.21$, with associated normalised eigenvectors $\widehat{e}_1, \widehat{e}_2, \widehat{e}_3$. $M = 1000$ points $y$ were uniformly drawn at random from the multivariate distribution Eq. (3.1), and projected onto the eigenvectors, defining the coordinates $z_i = \widehat{e}_i^T \cdot y$ with $i = 1, 2, 3$. The two panels locate the points in the $(z_1, z_2)$ (**B**) and $(z_2, z_3)$ (**C**) planes. The horizontal extension of the cloud of points in the top panel is about $\sqrt{\widehat{\lambda}_1/\widehat{\lambda}_2} \simeq 4.7$ times bigger than along the vertical axis, while both dispersions are comparable in the bottom panel ($\sqrt{\widehat{\lambda}_2/\widehat{\lambda}_3} \simeq 1.4$).

### 3.1.2 Most informative projection(s) and top component(s)

Let us try to apply the dimensional reduction idea to a configuration $y$ drawn from $p(y|\widehat{T})$ in Eq. (3.1). Suppose we cannot have access to $y$ directly, but can only measure its projection along some direction $v$ (with $|v|^2 = 1$):

$$x = v^T \cdot y + \xi , \tag{3.4}$$

where $\xi$ is a measurement noise, which we assume to be Gaussian, with zero mean and variance $\sigma_\xi^2$. How can we choose the direction of projection $v$ so that the scalar variable $x$ is maximally informative about the $L$-dimensional random variable $y$?

To answer this question, let us compute the mutual information between $x$ and $y$:

$$MI(x, y) = S[p(x)] - S[p(x|y)] , \tag{3.5}$$

which is the difference between the entropy of $x$ and its conditional entropy given $y$. Because $x$ is a linear combination of Gaussian variables, it is itself a Gaussian variable with zero mean and variance $\sigma_x^2 = v^T \cdot \widehat{C} \cdot v + \sigma_\xi^2$, where $\widehat{C}$ is the covariance matrix of $y$, see Eq. (3.3). The entropy of $x$ is thus given by

$$S[p(x)] = -\int dx \frac{e^{-x^2/(2\sigma_x^2)}}{\sqrt{2\pi\sigma_x^2}} \log\left(\frac{e^{-x^2/(2\sigma_x^2)}}{\sqrt{2\pi\sigma_x^2}}\right) = \frac{1}{2}\log\left(2\pi\,e\,\sigma_x^2\right) . \tag{3.6}$$

The entropy of $x$ conditioned to $y$ is simply the entropy of $\xi$, which is given by the same formula as above upon replacement of $\sigma_x^2$ with $\sigma_\xi^2$. Substracting the two entropies, we end up with

$$MI(x, y) = \frac{1}{2} \log \left( \frac{\sigma_x^2}{\sigma_\xi^2} \right) = \frac{1}{2} \log \left( 1 + \frac{v^T \cdot \widehat{C} \cdot v}{\sigma_\xi^2} \right) . \tag{3.7}$$

We can now select the best direction for the projection. The mutual information is maximal if we choose $v$ such that $v^T \cdot \widehat{C} \cdot v$ is maximal (under the normalisation constraint of the vector $v$). Therefore, the most informative direction is the eigenvector $\widehat{e}^1$ of $\widehat{C}$ associated to the largest eigenvalue $\widehat{\lambda}_1$. This eigenvector is called top component of $\widehat{C}$. Note that this result is independent of the value of the measurement noise variance. The role of $\xi$ is to ensure that the mutual information is well defined, and avoid the divergence in the entropy $S[p(x|y)]$ that would arise when the relation between the continuous variables $x$ and $y$ becomes deterministic.

This result is not surprising. If we want to compress a whole $L$-dimensional vector into a single scalar variable, we would rather have this variable to span as many values as possible to loose as little information as possible. An illustration can be found in figure 3.2. This reasoning can be straightforwardly extended to the case of multiple, say, $K$ projections. The most informative set of $K$ directions to be considered in order to maximise the mutual information between the measurements and the original variables are the top $K$ eigenvectors of $\widehat{C}$, namely, $\widehat{e}^1, \widehat{e}^2, ..., \widehat{e}^K$. This is the basic principle of PCA. Obviously, in real applications, one does not have access to the ground-truth covariance matrix $\widehat{C}$, but to some unbiased estimate computed from the data, such as the empirical covariance matrix with elements

$$C_{ij} = \frac{1}{M-1} \sum_{m=1}^{M} (y_i^m - \mu_i)(y_j^m - \mu_j) \quad \text{with} \quad \mu_i = \frac{1}{M} \sum_{m=1}^{M} y_i^m . \tag{3.8}$$

Here, the empirical averages $\mu_i$ are taken into account, as they generally do not vanish when $M$ is finite. What PCA amounts to in practice is to extract the top $K$ components of the empirical covariance matrix $C$ in Eq. (3.8), which we call $e^1, e^2, ..., e^K$. The important question is whether these vectors are good approximations of their ground-truth counterparts, $\widehat{e}^1, \widehat{e}^2, ..., \widehat{e}^K$.

### 3.1.3   The double $L, M \to \infty$ limit

The answer to this question undoubtedly depends on the amount of data available. If, for a fixed dimensionality $L$ of the configurations, the number $M$ of data grows, then $C$ will get closer and closer to $\widehat{C}$, and so will be the top components $e^1, e^2, ..., e^K$ to $\widehat{e}^1, \widehat{e}^2, ..., \widehat{e}^K$. More precisely, for a large but finite number of data, we expect the discrepancy between the empirical and ground-truth covariance due to imperfect sampling to scale as

$$C - \widehat{C} = \frac{Z}{\sqrt{M}} , \tag{3.9}$$

where the entries $Z_{ij}$ of the matrix $Z$ are positive or negative random variables of zero mean, whose variance converges to a finite limit equal to[1] $\widehat{C}_{ii}\widehat{C}_{jj} + \widehat{C}_{ij}^2$ in the limit $M \to \infty$.

We can use perturbation theory to determine the change in the eigenvectors due to these small perturbations on the matrix coefficients. For simplicity let us assume that the eigenvalues $\widehat{\lambda}^\ell$ of $\widehat{C}$ are non-degenerate. Perturbation theory tells us that, to the leading order in $1/\sqrt{M}$,

$$e_i^\ell - \widehat{e}_i^\ell \simeq \frac{1}{\sqrt{M}} \sum_{k(\neq\ell)} \frac{(\widehat{e}^k)^T \cdot Z \cdot \widehat{e}^\ell}{\widehat{\lambda}^k - \widehat{\lambda}^\ell} e_i^k . \tag{3.10}$$

Given the index $\ell$ of the component we look at, the sum over the other indices $k$ includes $L - 1$ terms. The numerator of the fraction is simply a matrix element of $Z$ in the eigenbasis of $\widehat{C}$, which is then of the order of unity, with zero mean. The denominator can be pretty small (for contiguous eigenvalues) but it will be of the order of the width of the spectrum of $\widehat{C}$ for most $k$, that is, of the order of unity too. Hence, we expect the sum in Eq. (3.10) to scale as $\sqrt{L}$ for large $L$.

As a conclusion, if $M \gg L$, we have enough data to estimate precisely the ground-truth top components from those of the empirical covariance matrix. On the contrary, if $M \ll L$, we cannot expect these top components to be reliably inferred from $C$. In between, if $L$ is comparable to $M$, the outcome of PCA is uncertain. We will see in the next section that there is a sharp transition between two regimes as $M$ and $L$ are sent to infinity at a fixed ratio

$$r = \frac{L}{M} . \tag{3.11}$$

The ratio $r$ between the sizes of the problem and of the dataset is a measure of the intensity of the noise due to sampling. If $r$ is smaller than some critical threshold $r^k$, then the $k^{th}$ top components of $C$ match, to some extent, those of $\widehat{C}$. Conversely, if $r > r^k$, $e^k$ will be practically orthogonal to $\widehat{e}^k$. The critical threshold $r^k$, whose value depends on the associated eigenvalue $\widehat{\lambda}^k$, is the slope of the dashed line appearing in figure 3.1, which separates the regions of possible and impossible inference of the relevant directions of variation in the data space.

## 3.2 The retarded learning phase transition

In this section we show the existence of the phase transition announced in section 3.1.3 when $r$ crosses its critical value. We do so for the simplest model of data, defining a specific direction $\widehat{e}$ in the data space.

Consider first the case of a Gaussian multivariate distribution where all $L$ variables are independent and have identical variances equal to one. The corresponding covariance and precision matrices, respectively, $C$ and $T$ are both equal to $\mathcal{I}_L$, the $L \times L$ identity matrix, see Eq. (3.2). As all eigenvalues of $C$ are equal to unity, the cloud of probability is isotropic, and no direction is privileged, as shown in figure 3.3**A**. In this

---

[1]We assume here again that $\boldsymbol{y}$ is drawn from the multivariate distribution Eq. (3.1).

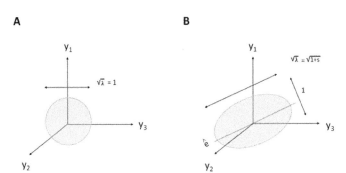

**Fig. 3.3** Schematic representation of multivariate Gaussian distributions in dimension $L = 3$. The grey zone corresponds to data points away from the mean (equal to zero) by less than one standard deviation. **A.** Case of independent variables, with $\widehat{T} = \mathcal{I}_L$. The cloud of probability is isotropic. **B.** Case of a precision matrix $\widehat{T}$ with one preferred direction $\widehat{e}$, see Eq. (3.13). The extension of the cloud along this axis is larger than along the orthogonal directions.

case, therefore, PCA does not carry any useful information, because all the directions are completely equivalent.

We then consider a second model, in which the cloud is made wider along one specific direction, say, $\widehat{e}$ (of unit norm). To do so, we focus on the multivariate Gaussian distribution associated to the covariance matrix

$$\widehat{C} = \mathcal{I}_L + \hat{s}\, \widehat{e} \cdot \widehat{e}^T \,, \tag{3.12}$$

where $\hat{s}$ is a positive real number. Note that $\widehat{e} \cdot \widehat{e}^T$ represents the projection operator onto the one-dimensional subspace along this specific direction. The corresponding cloud of probability is shown in figure 3.3**B**; it is elongated along $\widehat{e}$, with a width equal to the square root of the top eigenvalue $\widehat{\lambda} = 1 + \hat{s}$ of $\widehat{C}$. Along all the other, orthogonal directions the width of the cloud is equal to one. Note that the precision matrix corresponding to the covariance matrix above is

$$\widehat{T} = \widehat{C}^{-1} = \mathcal{I}_L - \frac{\hat{s}}{1 + \hat{s}}\, \widehat{e} \cdot \widehat{e}^T \,. \tag{3.13}$$

In the following, we consider this simplified, abstract model, and we discuss how and when the inference of $e$ is possible. Clearly, the value of $s$ is crucial here. If $s$ vanishes, no inference can be made because the direction $e$ does not pop out from the cloud in figure 3.3. The impossibility also extends to cases where $s$ is non-zero but "small" with respect to the sampling noise $r$. We will make this statement precise below. Of course, real data are often more structured, but the study of this very simplified model gives us some intuition on what we can expect in practical applications.

### 3.2.1   Bayesian justification for PCA

Suppose we draw $M$ independent configurations[2] encoded into a $M \times L$ matrix $Y = (y^1, y^2, \ldots, y^M)$ from the distribution in Eq. (3.1), where the precision matrix $\widehat{T}$ is

---

[2]Here, and in the following, we consider the data points $y$ as line vectors.

given by Eq. (3.13). Our goal is to infer the direction $\hat{e}$ defining $\hat{T}$ from the data $Y$. To do so, we write the log-likelihood (per data point) for a given precision matrix $T$:

$$\frac{1}{M} \log P(Y|T) = \frac{1}{M} \sum_{m=1}^{M} \log P(\boldsymbol{y}^m|T) \tag{3.14}$$

$$= \frac{1}{2} \log \det T - \frac{1}{2M} \sum_{m=1}^{M} (\boldsymbol{y}^m)^T \cdot T \cdot \boldsymbol{y}^m , \tag{3.15}$$

up to an irrelevant additive constant. As we are looking for a single direction $e$, we may use the same functional form for $T$ as in Eq. (3.13), with $e$ instead of $\hat{e}$, and $s$ instead of $\hat{s}$. Inserting this expression in the log-likelihood above we find

$$\frac{1}{M} \log P\left(Y|T = \mathcal{I}_L - \frac{s}{1+s} \, \boldsymbol{e} \cdot \boldsymbol{e}^T\right) = -\frac{1}{2} \log(1+s) + \frac{s}{2(1+s)} \boldsymbol{e}^T \cdot C \cdot \boldsymbol{e} , \tag{3.16}$$

where the matrix elements of $C$ are

$$C_{ij} = \frac{1}{M} \sum_{m=1}^{M} y_i^m y_j^m . \tag{3.17}$$

Note that we used the same notation for the matrix above as in Eq. (3.8). Because the empirical means $\mu_i$ vanish as $1/\sqrt{M}$, the difference between the matrices defined in Eqs. (3.8) and (3.17) is of the order of $1/M$, and does not affect any result in the double $L, M \to \infty$ limit. Later on, the matrix $C$ in Eq. (3.17) will be called indifferently empirical covariance matrix.

We now maximise the log-likelihood over $e$ (taking a uniform prior over the $(L-1)$-dimensional sphere $|e| = 1$) and over $s$. As a result our maximum *a posteriori* (MAP) estimator for $e$ is the normalised vector that maximises $\boldsymbol{e}^T \cdot C \cdot \boldsymbol{e}$, that is, the eigenvector associated to the top eigenvalue of $C$, which we call $\lambda_{max}$. Hence, the Bayesian inference framework justifies the use of PCA in the case of multivariate Gaussian distributions. Of course PCA, intended as the search for the top component of the empirical correlation matrix, can be used in practice to analyse all kind of data, even unrelated to any Gaussian distributions, but with no guarantee that the method will provide relevant directions in the data space.

### 3.2.2 Largest eigenvalue of the empirical correlation matrix

We are left with the characterisation of the top eigenvalue and eigenvector of the empirical covariance matrix $C$ in Eq. (3.17). Two important remarks will help:

- For any invertible matrix $P$, the spectrum of $P \cdot C \cdot P^{-1}$ is the same as that of $C$. Because $\hat{T}$ is definite positive, it is invertible, and so are all its powers (in the matrix sense). Choosing $P = \hat{T}^{1/2}$, we obtain

$$\text{Spectrum}(C) = \text{Spectrum}(\hat{T}^{1/2} \cdot C \cdot \hat{T}^{-1/2}) = \text{Spectrum}\left(\hat{T}^{1/2} \cdot \frac{Y^T \cdot Y}{M} \cdot \hat{T}^{-1/2}\right)$$

$$= \text{Spectrum}\left(\frac{X^T \cdot X}{M} \cdot \hat{C}\right) \quad \text{where} \quad X = Y \cdot \hat{T}^{1/2} , \tag{3.18}$$

and we have used $\widehat{C} = \widehat{T}^{-1}$.

- Each data point in the "new" dataset $X = (x^1, x^2, ..., x^M)$ is defined, according to the equation above, as a linear transform of the corresponding "old" data point, $x^m = y^m \cdot \widehat{T}^{1/2}$. Inserting this change of variable in the multivariate Gaussian distribution for $y$, see Eq. (3.1), we immediately see that $x$ is itself a multivariate Gaussian variable, associated to the precision matrix $\widehat{T}_{xx} = \mathcal{I}_L$. In other words, the components $x_i$ of $x$ are independent Gaussian variables, with zero mean and unit variance.

As a result,

$$C_{xx} = \frac{1}{M} X \cdot X^T \tag{3.19}$$

is the empirical covariance matrix computed from a set of $M$ configurations of $L$ independent normal variables. Such matrices were introduced by Wishart a century ago [12]. As there is no correlation between the variables, when $M \gg L$, all eigenvalues $\lambda$ tend to one. However, for a finite aspect ratio $r$, as defined in Eq. (3.11), of the data matrix $X$, the spectrum of $C_{xx}$ is non trivial, with a support extending on a portion of the real positive axis. We will discuss the typical spectral properties of $C_{xx}$ in section 3.2.3.

Based on the discussion above, let $z$ be an eigenvalue of $C$. Then $z$ is also an eigenvalue of $C_{xx} \cdot \widehat{C}$, and we write

$$0 = \det(z\,\mathcal{I}_L - C_{xx} \cdot \widehat{C}) = \det\left(z\,\mathcal{I}_L - C_{xx} - \hat{s}\,C_{xx} \cdot \widehat{e} \cdot \widehat{e}^T\right) . \tag{3.20}$$

We are left with two possibilities. Either $z$ is in the spectrum of $C_{xx}$, which we shall discuss in the next section, or $z\,\mathcal{I}_L - C_{xx}$ is invertible, and, according to Eq. (3.20) and the general equality $\det(A \cdot B) = \det A \times \det B$,

$$0 = \det\left(z\,\mathcal{I}_L - C_{xx} - \hat{s}\,C_{xx} \cdot \widehat{e} \cdot \widehat{e}^T\right) = \det\left(\mathcal{I}_L - \hat{s}\,\mathcal{M}\right) , \tag{3.21}$$

where

$$\mathcal{M} = (z\,\mathcal{I}_L - C_{xx})^{-1} \cdot C_{xx} \cdot \widehat{e} \cdot \widehat{e}^T . \tag{3.22}$$

Due to the presence of the projector onto $e$ in the expression of $\mathcal{M}$, the image of any vector orthogonal to $e$ through $\mathcal{M}$ vanishes. Hence $\mathcal{M}$ is of rank 1, and its unique non-zero eigenvalue is equal to its trace. Eq. (3.21) is therefore equivalent to

$$\widehat{e}^T \cdot (z\,\mathcal{I}_L - C_{xx})^{-1} \cdot C_{xx} \cdot \widehat{e} = \frac{1}{\hat{s}} . \tag{3.23}$$

To solve this implicit equation over $z$ we need first to discuss the spectral properties of $C_{xx}$.

### 3.2.3    Marcenko-Pastur spectrum of random covariance matrices

Wishart random matrices have been the subject of intensive studies over past decades, and are now well characterised, see *e.g.* [13–15]. In particular, the spectrum of $C_{xx}$ has been derived by Marcenko and Pastur (MP) [16] in the double $L, M \to \infty$ limit at fixed ratio $r = L/M$, with the following results:

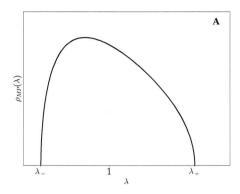 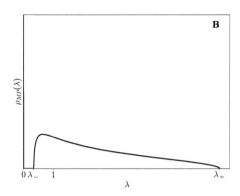

**Fig. 3.4** Marcenko-Pastur densities of eigenvalues $\rho(\lambda)$ vs. $\lambda$ for sampling ratios $r = 0.05$ (**A**) and $r = 2.5$ (**B**). Note the presence of the Dirac distribution in $\lambda = 0$ when $r > 1$, see Eq. (3.26).

- The distribution of the eigenvalues of $C_{xx}$ is self-averaging, *i.e.* does not depend on the particular realisation of the data $X$ in this large size limit. Deviations are possible, but happen with vanishing probability when $L, M \to \infty$. We will come back to this important point below.
- In the double infinite size limit, the density of eigenvalues becomes continuous, and is supported by the interval $\lambda \in [\lambda_-, \lambda_+]$, where

$$\lambda_- = (1 - \sqrt{r})^2 \quad \text{and} \quad \lambda_+ = (1 + \sqrt{r})^2 . \tag{3.24}$$

In particular, we see that all eigenvalues concentrate in $\lambda = 1$ for perfect sampling ($r = 0$). For finite $r$, the width of the spectrum is finite, and increases for poorer sampling (larger $r$).

- For good sampling, *i.e.* $r < 1$, the density of eigenvalues is given by

$$\rho_{MP}(\lambda) = \frac{\sqrt{(\lambda_+ - \lambda)(\lambda - \lambda_-)}}{2\pi r \lambda} , \qquad r < 1 . \tag{3.25}$$

It vanishes at the edges of the support, $\lambda = \lambda_\pm$, and is strictly positive in between[3]. Its maximum is reached in $\lambda = (1 - r)^2/(1 + r)$.

- For poor sampling, *i.e.* $r > 1$, Eq. (3.25) still gives the continuous distribution of eigenvalues, but another contribution must be considered. Because the rank of $C_{xx}$ is at most $M$, $L - M$ eigenvalues are equal to zero, giving a Dirac distribution in $\lambda = 0$ with associated weight $1 - 1/r$. The resulting distribution is therefore

$$\rho_{MP}(\lambda) = \frac{\sqrt{(\lambda_+ - \lambda)(\lambda - \lambda_-)}}{2\pi r \lambda} + \left(1 - \frac{1}{r}\right) \delta(\lambda) , \qquad r > 1 . \tag{3.26}$$

---

[3]When $r \ll 1$ the density of eigenvalues becomes, to the leading order in $r$, a semi-circle law of width $2\sqrt{r}$, centred around $\lambda = 1$. The reason is that, when $M \gg L$, the correlations between the different entries of $C_{xx}$ can be neglected, and the empirical covariance matrix is essentially a random Gaussian matrix.

These results are graphically summarised in figure 3.4. A pedagogical analytical derivation of the above density of eigenvalues can be found *e.g.* in Ref. [14].

Let us remark that the lack of structure in the data $X$ has some consequences on the nature of the eigenvectors of $C_{xx}$. The mean and variance of a component of any one of its eigenvectors, say, $u_i$, are equal to, respectively, zero and $1/L$ (due to normalisation). In agreement with the maximum entropy principle, the distribution of $u_i$ is Gaussian, with zero mean and unit variance. The joint distribution of any (finite compared to $L, M$) number of eigenvector components is a product of such Gaussian laws, see *e.g.* [15].

The Marcenko-Pastur distribution is important in statistics, as it provides a null model for assessing the relevance of PCA. In other words, it gives the natural width of the eigenvalue distribution associated to a structureless null model (with precision matrix equal to the identity), in presence of undersampling, *i.e.* when the number of data points $M$ is of the same order of $L$. If the top eigenvalue of the empirical covariance matrix of some dataset one is dealing with does not exceed $\lambda_+$, it is likely that the corresponding eigenmode is not capturing any structure in the data, but is rather due to undersampling, *i.e.* it is compatible with the trivial null model. On the contrary, if the top eigenvalue is much larger than the edge $\lambda_+$, then it cannot be explained by mere noise in the data, and must convey some statistically robust signal. As we will see below, this reasoning is qualitatively correct, but does not capture the right relationship between $s$ and $r$ at the critical threshold for reconstruction[4].

### 3.2.4   The phase transition

We are now ready to conclude the discussion of the spectrum of the empirical covariance matrix $C$. As we have seen in section 3.2.2, the eigenvalues $z$ of $C$ can be either the same as those of $C_{xx}$ and lie in the interval $[\lambda_-, \lambda_+]$, see Eq. (3.24), or the root of the implicit Eq. (3.23).

To solve the latter equation we introduce the eigenvalues $\lambda^\alpha$ and eigenvectors $\boldsymbol{u}^\alpha$ of $C_{xx}$, with $\alpha = 1, \dots, L$. Eq. (3.23) becomes

$$\sum_{\alpha=1}^{L} \frac{\lambda_\alpha}{z - \lambda_\alpha} \left(\boldsymbol{e}^T \cdot \boldsymbol{u}^\alpha\right)^2 = \frac{1}{\hat{s}} \ . \tag{3.27}$$

Consider now a small interval of eigenvalues in the support of the Marcenko-Pastur spectrum, say, $[\lambda, \lambda + \epsilon]$. The number of eigenvalues falling in this interval is approximately $L \times \rho_{MP}(\lambda) \times \epsilon$. Let us now choose $\epsilon \to 0$ such that $L \times \epsilon \to \infty$, *e.g.* $\epsilon \sim L^{-1/4}$: such a choice ensures that all the eigenvalues in the interval take almost the same value $\lambda$, yet their number diverges. The contribution of the modes $\alpha$ falling into this interval to the sum in Eq. (3.27) reads

$$\sum_{\alpha:\lambda<\lambda^\alpha<\lambda+\epsilon} \frac{\lambda_\alpha}{z - \lambda_\alpha} \left(\boldsymbol{e}^T \cdot \boldsymbol{u}^\alpha\right)^2 \simeq \frac{\lambda}{z - \lambda} \sum_{\alpha:\lambda<\lambda^\alpha<\lambda+\epsilon} \left(\boldsymbol{e}^T \cdot \boldsymbol{u}^\alpha\right)^2 \ . \tag{3.28}$$

---

[4]Taken literally, this reasoning would lead to the statement that reconstruction of the top eigenvector is possible for $s$ larger than a threshold given by the relation $1+s = (1+\sqrt{r})^2$, *i.e.* $s = 2\sqrt{r}+r$. We will see in the next section that the correct relation is $s = \sqrt{r}$.

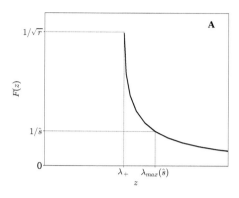 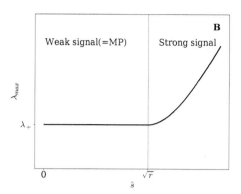

**Fig. 3.5** Top eigenvalue of the empirical correlation matrix. **A.** The function $F(z)$ and the graphical resolution of Eq. (3.29) for the maximal eigenvalue $\lambda_{max}$ of the empirical covariance matrix. **B.** Plot of $\lambda_{max}$ and phase diagram for the retarded learning transition. If the signal $\hat{s}$ is smaller than $\sqrt{r}$, learning of the component **e** is impossible (weak signal regime). Otherwise, the top component of the empirical correlation matrix has a non-zero dot product with **e** (strong signal regime).

The quantity $\mathbf{e}^T \cdot \mathbf{u}^\alpha$ is the projection of the $\alpha^{th}$ eigenvector onto **e**. As discussed in the previous section, we expect this projection to be distributed as a Gaussian, with zero mean and variance equal to $1/L$. The sum of a diverging number of the squared projections will then converge towards their variance times the number of terms in the sum, that is, $\rho_{MP}(\lambda) \times \epsilon$.

Summing over all intervals of width $\epsilon \to 0$ gives us the following implicit equation for the eigenvalue $z$:

$$F(z) \equiv \int_{\lambda_-}^{\lambda_+} d\lambda \, \rho_{MP}(\lambda) \, \frac{\lambda}{z - \lambda} = \frac{1}{\hat{s}} \, . \tag{3.29}$$

Note that the $\delta(\lambda)$ term that is present in $\rho_{MP}(\lambda)$ for $r > 1$ does not contribute because the rest of the integrand vanishes at $\lambda = 0$. We look for a solution for $z$ outside of the support of $\rho_{MP}$, more precisely, larger than $\lambda_+$.

The integral in Eq. (3.29) can be computed exactly from the MP expression for the density, see Eq. (3.25), and is related to the so-called Stieltjes transform of $\rho_{MP}$, a key object in random matrix theory [14]. We obtain

$$F(z) = \frac{1}{2r} \left( z - r - 1 - \sqrt{z^2 - 2z(1 + r) + (1 - r)^2} \right) , \qquad z > \lambda_+ \, . \tag{3.30}$$

The behaviour of $F(z)$ is sketched in figure 3.5**A**. As $z$ decreases from $+\infty$, $F(z)$ increases, and reaches its maximum $F = \frac{1}{\sqrt{r}}$ for $z = \lambda_+$. We can now conclude:

- if $r < r_c = \hat{s}^2$, there exists a unique $z > \lambda_+$ solution of Eq. (3.30). As all the other eigenvalues are lower than $\lambda_+$, $z$ coincides with $\lambda_{max}$, the top eigenvalue of $C$. Its expression is

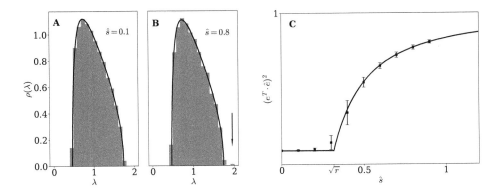

**Fig. 3.6** Illustration of the retarded learning phase transition. **A, B**. Histograms of eigenvalues obtained for two representative realisations of the empirical covariance matrix with $L = 1000$ variables and $r = 0.1$, for $\hat{s} = 0.1$ (**A**) and $\hat{s} = 0.8$ (**B**). The two histograms are in very good agreement with the Marcenko-Pastur distribution (full line), except for the presence of a large signal eigenvalue located by the arrow for $\hat{s} = 0.8$. **C**. Squared overlap between the ground truth and the empirical top component as a function of $\hat{s}$. The full line shows Eq. (3.33) for $r = 0.1$, while the points with error bars are results of numerical simulations with $L = 1000$ and $M = 10000$, averaged over 100 samples.

$$\lambda_{max} = 1 + \hat{s} + r + \frac{r}{\hat{s}} , \qquad r < \hat{s}^2 . \tag{3.31}$$

Note that $\lambda_{max}$ is larger than the maximal eigenvalue of $\widehat{C}$, $1+\hat{s}$, as soon as $r > 0$.

- if $r > r_c = \hat{s}^2$, then Eq. (3.30) does not admit any solution $z \geq \lambda_+$. Hence all the eigenvalues of $C$ coincide with those of $C_{xx}$, with a density given by the MP expression. The largest eigenvalue of $C$ is thus

$$\lambda_{max} = (1 + \sqrt{r})^2 , \qquad r > \hat{s}^2 . \tag{3.32}$$

The behaviour of the maximal eigenvalue of $C$ as a function of $\hat{s}$ is shown in figure 3.5B; it is equal to the right edge of the MP spectrum $\lambda_+$ for $r \geq r_c$, and pops out from this noisy spectrum for smaller $r$ or larger $s$. For excellent sampling $r \to 0$, it reaches $1+\hat{s}$, *i.e.* the top eigenvalue of $\widehat{C}$. It is important to stress that, while the above derivation is correct in the double $L, M \to \infty$ limit, the two regimes can also be observed for finite sizes, see figure 3.6**A, B**.

The phenomenon taking place at $r = r_c$ is a true phase transition, separating a phase where sampling is insufficient to learn the relevant direction $e$ from a phase where $e$ can be inferred. It is an example of a general phenomenon, called retarded learning [17]. Retarded learning means that, contrary to intuition, the inference of the model parameters (here, the direction $\hat{e}$) is not happening as soon as some data are available, but requires a *critical amount* of data (here, $M = L/\hat{s}^2$) to take place. The squared dot product between the top components of $C$ and $\widehat{C}$ can be computed as a function of $r, \hat{s}$ [18]. It vanishes for $r > r_c$, showing the failure of PCA to find the direction $\hat{e}$, and is positive at smaller ratios,

$$\left( e^T \cdot \hat{e} \right)^2 = \frac{1 - r/\hat{s}^2}{1 + r/\hat{s}} \ , \qquad r < \hat{s}^2 \ , \qquad (3.33)$$

see figure 3.6**C**.

To end with, let us emphasise that the results and discussion above remain valid if $K \geq 2$ (finite when $L, M \to \infty$) components $\hat{e}_1, \hat{e}_2, ..., \hat{e}_K$ are associated to eigenvalues larger than one, say, $1 + \hat{s}_1, 1 + \hat{s}_2, ..., 1 + \hat{s}_K$, with $\hat{s}_1 \geq \hat{s}_2 \geq ... \geq \hat{s}_K$. As more and more data are acquired and the sampling ratio $r$ decreases and successively crosses the critical values $r_c^1 = \hat{s}_1^2$, then $r_c^2 = \hat{s}_2^2$, and so on until $r_c^k = \hat{s}_K^2$ , the corresponding components are inferred one after the other.

In summary, we have seen that there is a fundamental difference between low-dimensional and high-dimensional inference. In the low-dimensional case, one continuously acquires information about the unknown parameters when the number of available data is increased. Furthermore, when $M \to \infty$ convergence of the inferred parameters to the ground truth is guaranteed, unless a very constraining prior is imposed. Conversely, in the high-dimensional case, a phase transition separates a regime where inference is impossible from one where inference is possible. When both $M, L \to \infty$, for $M < L/r_c$ no information on the ground truth can be obtained from the data, while for $M > L/r_c$ some information can be extracted. We will further discuss retarded learning in Chapter 7.

## 3.3  Tutorial 3: replay of neural activity during sleep following task learning

Understanding how memories are stored and consolidated in the brain is one of the fundamental goals of neuroscience. In mammals, a brain area called hippocampus is known to be crucial for the formation and acquisition of short-term memories. At later times, these memories are transformed into long-term memories and stored in the prefrontal cortex. It is important to better understand how memories are transferred from the hippocampus to the prefrontal cortex, and the role of sleep in this memory consolidation process. To such aim, multi-electrode recording of a few dozens of neurons in the prefrontal cortex of alive rats were performed in [19, 20]. A single recording session, in which the same neurons are recorded, is divided into three epochs: a *task* epoch in which a rat has to learn, through trials and errors, what is the rewarded arm in a Y-shaped maze (there are four possible rules that are changed by the operator as soon as the rat learns them: go to the left arm, go to the right arm, go where a light is on, go where a light is off); and two *sleep* epochs, one before and the other after the task epoch. The goal of this tutorial is to analyse the recorded data using principal component analysis and tools from random matrix theory, in order to detect similarities between the neuronal activity during the task and post-task sleep epochs, which where not present in the pre-task sleep epoch.

### 3.3.1  Problem

During a recording session, $L$ neurons, labelled by $i = 1, \ldots, L$, are recorded for a total time $T$. Following Refs. [19, 20], the spike train has been discretised in $\Delta t = 100 \, ms$ time bin. We call $M = T/\Delta t$ the total number of bins, and $t_b$ the time corresponding to the centre of the $b^{th}$ bin, $b = 1, \ldots, M$. The activity of a neuron $i$ in a time bin $[t_b - \Delta t/2, t_b + \Delta t/2]$ is represented by a spike count variable $s_i(t_b) = 0, 1, 2, \ldots$. Note that for small time bin (*e.g.* $\Delta t = 10$ ms), this is equivalent to the binary representation used in tutorial 2.

**Pearson correlation matrix:** We define the mean and variance of $s_i$ over time as

$$p_i = \langle s_i \rangle = \frac{1}{M} \sum_{b=1}^{M} s_i(t_b) ,$$

$$\sigma_i^2 = \frac{1}{M-1} \sum_{b=1}^{M} \left[ s_i(t_b) - p_i \right]^2 . \tag{3.34}$$

The *z-score* variables, which will be our data, are defined as

$$y_{bi} = \frac{s_i(t_b) - p_i}{\sigma_i} , \tag{3.35}$$

which defines a $M \times L$ matrix $Y$. Each line $\boldsymbol{y}_i$ of the matrix corresponds to a realisation of the spiking process of the $L$ neurons, and we have $M$ such realisations available, each corresponding to a different time. We define the $L \times L$ Pearson correlation matrix as:

$$C = \frac{1}{M} Y^T Y \qquad \text{or} \qquad C_{ij} = \frac{1}{M} \sum_{b=1}^{M} y_{bi} y_{bj} \ . \qquad (3.36)$$

The matrix element $C_{ij}$ encodes the correlation, averaged over time, of a pair of neurons $i, j$. Such matrix can be diagonalised to find its eigenvectors and eigenvalues. The central part of the spectrum of the eigenvalues is due to sampling noise, and therefore it follows a Marcenko-Pastur law of parameter $r = L/M < 1$, see Eq. (3.24). According to the discussion of Chapter 3, in order to extract the signal components, we must select the eigenvalues that fall outside the interval $[\lambda_-, \lambda_+]$.

**Replay:** We want to identify the similarity in the correlation structure between the *sleep* neuronal activity (in the *pre-task* or *post-task* epochs) and the *task* activity. Following Peyrache et al. [20], the task epoch is called the *template* epoch, while the *pre-task* or *post-task* sleep epochs are called *match* epochs. An instantaneous measure of similarity in the correlated activity is obtained by computing the scalar product, over all the pairs of neurons, of the simultaneous co-activation of the pair during the match and the corresponding pair correlation during the template,

$$R^{match}(t_b) = \frac{1}{2} \sum_{i \neq j} y_{bi}^{match} C_{ij}^{template} y_{bj}^{match} \ , \qquad (3.37)$$

which, after averaging over the time bins, gives the total replay:

$$R_T^{match} = \frac{1}{2} \sum_{i \neq j} C_{ij}^{match} C_{ij}^{template} \approx \frac{1}{2} \text{Tr} \left[ (C^{match} - \mathcal{I}_L)(C^{template} - \mathcal{I}_L) \right], \qquad (3.38)$$

where Tr is the trace operation, $\mathcal{I}_L$ is the $L \times L$ identity matrix, and we used that, by definition, $C_{ii} = (M-1)/M \sim 1$. Using the eigenvalue decomposition $C_{ij}^{template} = \sum_k \lambda_k v_i^k v_j^k$ in Eq. (3.37), one obtains the time-dependent replay as a sum of replays associated to each principal component:

$$R^{match}(t_b) = \sum_k \lambda_k R_k^{match}(t_b) \ , \qquad (3.39)$$

where

$$R_k^{match}(t_b) = \frac{1}{2} \left[ \left( \sum_i y_{bi}^{match} v_i^k \right)^2 - \sum_i (y_{bi}^{match} v_i^k)^2 \right] \ , \qquad (3.40)$$

keeping in mind that $v^k$ are the eigenvectors associated to the template epoch.

**Data:**
The spiking activity of a population of $L = 37$ prefrontal cortex neurons during the *pre-task* sleep (*Dati_181014_sleep1.txt*), task (*Dati_181014_maze.txt*) and *post-task* sleep (*Dati_181014_sleep2.txt*) epochs can be downloaded from the book webpage[5], see tutorial 3 repository. Each file contains a two-column array $\{t_i, i\}$, where $t_i$ is a spiking

---

[5]`https://github.com/StatPhys2DataDrivenModel/DDM_Book_Tutorials`. The original data can be found on `https://crcns.org/data-sets/pfc/pfc-6`.

time of neuron $i$ (times are in units of 0.1 ms), ranging between $T_{min}$ and $T_{max}$, and $i$ is the neuron index. Data are taken from Ref. [19].

**Questions:**

In the tutorial 3 data repository, you are given a Jupyter notebook *tut3_start.ipynb* that reads the data and answers to questions 1 and 2. You can use it to start working more quickly. Complete the notebook or write your own code to answer the following questions.

1. Read the data and produce the raster plot for the $L$ neurons and the first 10 seconds of the activity. Note that in reading the data, you should find $T_{min}$ and $T_{max}$ for each epoch.

2. Compute the Pearson correlation matrix of the *task* activity and diagonalize it. Compare the spectrum of eigenvalues of $C^{task}$ to the Marcenko-Pastur distribution to extract the signal components associated to large eigenvalues. How many signal components with $\lambda > \lambda_+$ are present?

3. Reshuffle the data, by randomly permuting the activity of the single neurons over the $M$ possible time bins, in such a way that the correlations between neurons spiking together are lost, but the frequencies of individual neurons are kept constant. Verify that the spectrum of the eigenvalues of the Pearson correlation matrix of the shuffled data is well described by the Marcenko-Pastur distribution.

4. Identify the neurons that contribute the most to the largest component of the *task* activity. These neurons identify a cell assembly of neurons coactivating during the task.

5. Calculate the total replay of the *pre-task* sleep and *post-task* sleep epochs, with respect to the *task* epoch.

6. Plot the replay as a function of time of the first component in the *pre-task* and *post-task* sleep epochs. Is it constant in time? Compare with the results in Ref. [20].

7. Compare the total reactivation of the first and second components. What components are replayed in the *post-task* sleep epoch?

### 3.3.2 Solution

The programme *tutorial3.ipynb* reads the data of the *task* activity and compares it to the *pre-task* and *post-task* sleep activity. Following Ref. [20], the time bin $\Delta t$ is fixed to $\Delta t = 100$ ms.

1. The raster plot of data for the 37 neurons in the *task* epoch is shown in figure 3.7 for the first 10 seconds of recording.

2. The eigenvalue spectrum of the Pearson correlation matrix in the *task* epoch is shown in figure 3.8**A** and compared to the Marcenko-Pastur distribution in Eq. (3.24)

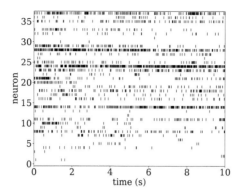

**Fig. 3.7** Raster plot of the activity of the $L = 37$ neurons in the *task* epoch. Only the first 10 seconds of the recording are shown.

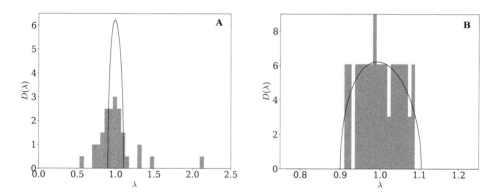

**Fig. 3.8** **A**. Spectrum of the eigenvalues of the Pearson correlation matrix in the *task* epoch, compared with the Marcenko-Pastur distribution. **B**. Spectrum of the eigenvalues of the Pearson correlation matrix of the reshuffled activity in the *task* epoch, compared with the Marcenko-Pastur distribution. The figure agrees with figure 1 in Ref. [20] (the session analysed here corresponds to the last line of the figure).

with $r = L/M$ and $\lambda_\pm = (1 \pm \sqrt{r})^2$. There are six principal components defined as signal with $\lambda > \lambda_+$.

3. The spectrum of the reshuffled spiking activity (no correlations between variables) is in perfect agreement with the Marcenko-Pastur distribution, as shown in figure 3.8**B**.

4. As shown in figure 3.9, the neurons that mostly contribute to the first component of the Pearson correlation matrix in the *task* epoch are those labelled as $\{1, 9, 20, 21, 26, 27\}$. This is the cell assembly that is replayed in the *post-task* sleep and is therefore associated with the learning of the task.

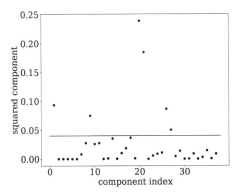

**Fig. 3.9** Squared entries of the first principal component of the Pearson correlation matrix of the activity in the *task* epoch. The six largest neuron components, corresponding to neurons labelled as $\{1, 9, 20, 21, 26, 27\}$ are above the horizontal line.

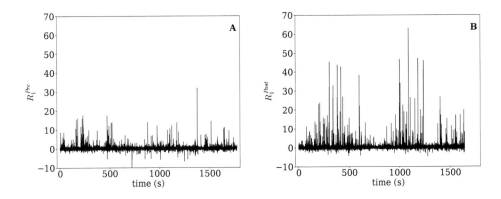

**Fig. 3.10** Reactivation as function of the time of the first principal component of the *task* epoch in the *pre-task* (left) and *post-task* (right) sleep epochs. This figure is in agreement with figure 2 in Ref. [20] (note that our definition of $R$ is half that of Ref. [20]).

5. The total replay in the *pre-task* and *post-task* sleep epochs are, respectively, 0.29 and 0.55. The fact that the replay in the *pre-task* sleep epoch is significatively smaller than that in the *post-task* sleep epoch suggests that the higher replay in the *post-task* sleep is associated with the learning process during the *task* epoch.

6. The behaviour as a function of time of the reactivation of the first component of the *task* epoch in the *pre-task* and *post-task* sleep epochs are shown in figure 3.10. The reactivation in the *post-task* sleep epoch is not constant over time, but it is concentrated in a particular period of the sleep, called non-REM, and in particular when inputs coming from the hippocampus called sharp-waves arrive in the prefrontal cortex.

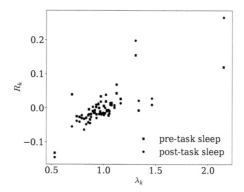

**Fig. 3.11** Component replay $R_k$ in the *pre-task* (circles) and *post-task* (squares) epochs versus eigenvalue $\lambda_k$. Note that the total replay is $\sum_k \lambda_k R_k$, see Eq. (3.39).

7. The total replay of the first component in the *post-task* sleep epoch is 0.27. The missing contribution does not come from the following signal component: the total contribution of the second component to the replay is indeed 0.03. It is shown in figure 3.11 that the missing contribution come mostly from the fourth component, and from signal components with $\lambda < \lambda_-$.

# 4

# Priors, regularisation, sparsity

So far we have not discussed much the role of the prior distribution $p(\boldsymbol{\theta})$ in Bayes' rule,

$$p(\boldsymbol{\theta}|\boldsymbol{y}) = \frac{p(\boldsymbol{y}|\boldsymbol{\theta})\,p(\boldsymbol{\theta})}{p(\boldsymbol{y})}\;. \tag{4.1}$$

To be more precise we have considered priors $p(\boldsymbol{\theta})$ that did not depend on $\boldsymbol{\theta}$, the so-called uniform priors. There are, however, several reasons to avoid this hypothesis:

- A uniform prior is not optimal in many situations. We may actually have some knowledge about $\boldsymbol{\theta}$, and want to exploit it. Conversely, when a complex model has to be inferred from a limited set of data, use of a prior can help reduce the effective size of the parameter space and avoid overfitting the data. As an illustration, the critical sampling ratio $r$ locating the retarded learning phase transition in principal component analysis (discussed in Chapter 3) can be increased with appropriate choices of priors over the entries of the top component [21, 22].
- Adequate priors are needed when the likelihood alone does not define a well-conditioned or well-behaved posterior distribution. We have seen this issue in the German tank problem: with a uniform prior high, order moments of the maximal tank number are not defined.
- Furthermore, uniform priors may not be logically consistent. What is uniform for some variable $\boldsymbol{\theta}$ is no longer uniform after a change of variable $\boldsymbol{\theta} \to \boldsymbol{\theta}'$ in the model definition. Is there a non-arbitrary choice of priors that would be insensitive to such reparametrisations?

In this chapter we will discuss three classes of priors that are useful to address the three issues above: $L_p$-norm based priors, conjugate priors and invariant priors.

## 4.1  $L_p$-norm based priors

$L_p$-norm based priors are very useful in many practical applications. The idea is to favour, in the course of the inference, parameter vectors $\boldsymbol{\theta}$ with small $L_p$ norms,

$$\left|\boldsymbol{\theta}\right|_p = \left(\sum_{d=1}^{D} \theta_d^p\right)^{1/p}, \tag{4.2}$$

by choosing for the prior distribution,

$$p(\boldsymbol{\theta}) \propto \exp\left(-\frac{\gamma}{p!}\left|\boldsymbol{\theta}\right|_p^p\right) = \exp\left(-\frac{\gamma}{p!}\sum_{d=1}^{D}\theta_d^p\right), \tag{4.3}$$

where $\gamma > 0$ is the strength of the prior, and we have omitted the normalisation constant[1]. $L_p$-norm based priors can be thought as regularisers: they penalise large parameter values that may arise through maximisation of the log-likelihood. For large $\gamma$, the components of $\boldsymbol{\theta}$ are strongly pushed to be small, which is an effective way to reduce their number. Besides, the value of $p$ qualitatively affects how the parameters are "pushed" towards zero, as we will see below.

### 4.1.1 Maximum likelihood estimator for least-square regression

We begin by illustrating the action of $L_p$ priors in a simple albeit important case: least-square regression. Let us assume we have a $L$-dimensional data point, $\boldsymbol{y}$, that linearly depends on $D$ unknown parameters, $\boldsymbol{\theta}$, through a known matrix $A$ of dimensions $L \times D$,

$$\boldsymbol{y} = A \cdot \boldsymbol{\theta} + \boldsymbol{z}, \tag{4.4}$$

where $\boldsymbol{z}$ is a Gaussian measurement noise with zero mean and unit variance (uncorrelated from component to component). The logarithm of the likelihood reads

$$\log p(\boldsymbol{y}|\boldsymbol{\theta}) = -\frac{1}{2} \sum_{i=1}^{L} \left( y_i - \sum_{d=1}^{D} A_{i,d} \theta_d \right)^2, \tag{4.5}$$

up to an irrelevant additive constant. The maximum likelihood estimator (MLE) is obtained by maximising this quantity over $\boldsymbol{\theta}$. Differentiation with respect to $\theta_d$ gives

$$\frac{\partial}{\partial \theta_d} \log p(\boldsymbol{y}|\boldsymbol{\theta}) \bigg|_{\boldsymbol{\theta}^{MLE}} = 0 = \sum_{i=1}^{L} \left( y_i - \sum_{d'=1}^{D} A_{i,d'} \theta_{d'}^{MLE} \right) A_{i,d} \tag{4.6}$$

for all $d = 1, \dots, D$ or, equivalently, in vectorial notations,

$$A^T \cdot A \cdot \boldsymbol{\theta}^{MLE} = A^T \cdot \boldsymbol{y}. \tag{4.7}$$

If $L \geq D$ we can generically deduce $\boldsymbol{\theta}$ from the above equation, with the unique solution

$$\boldsymbol{\theta}^{MLE} = (A^T \cdot A)^{-1} \cdot A^T \cdot \boldsymbol{y}. \tag{4.8}$$

For $L < D$, however, the matrix $A^T \cdot A$ is not invertible, and there exist infinitely many solutions to Eq. (4.7). By choosing a prior we will lift this indeterminacy.

### 4.1.2 $L_2$-norm prior: ridge regression

In the $p = 2$ case, the $L_2$ prior is Gaussian:

$$p(\boldsymbol{\theta}) = \prod_{d=1}^{D} \frac{e^{-\frac{\gamma}{2}\theta_d^2}}{\sqrt{2\pi/\gamma}}. \tag{4.9}$$

For the linear dependence defined by Eq. (4.4), MLE with a $L_2$ prior is also called ridge regression. It is easy to add the logarithm of this prior to the log-likelihood associated

---

[1]The term $p!$ in Eq. (4.3) can be replaced by $\Gamma(p+1)$ for non-integer $p$.

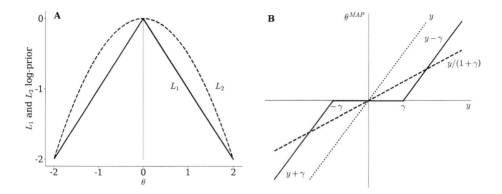

**Fig. 4.1  A.** Logarithm of the priors corresponding to the $L_1$ (full line) and $L_2$ (dashed line) norms. Here, $\gamma = 1$ in both cases. **B.** Maximum a posteriori value of the parameter, $\theta^{MAP}$, when no regularisation is applied (dotted line), using a $L_1$-norm prior as in Eq. (4.14) (full line), and using a $L_2$-norm prior as in Eq. (4.15) (dashed line).

to least-squared regression in Eq. (4.5), and maximise over $\boldsymbol{\theta}$ the corresponding log-posterior. The result is the maximum a posteriori (MAP) estimator of the parameters,

$$\boldsymbol{\theta}_2^{MAP} = \left(A^T \cdot A + \gamma \mathcal{I}_D\right)^{-1} \cdot A^T \cdot \boldsymbol{y} \ . \tag{4.10}$$

This expression is now well defined for all $L, D$, because $A^T \cdot A$ is a non-negative matrix. Note that the two limits

$$\lim_{\gamma \to 0^+} \left(A^T \cdot A + \gamma \mathcal{I}_D\right)^{-1} \cdot A^T = \lim_{\gamma \to 0^+} A^T \cdot \left(A \cdot A^T + \gamma \mathcal{I}_L\right)^{-1} \tag{4.11}$$

exist and coincide; this common limit defines a $D \times L$ matrix called the pseudo-inverse of $A$, a generalisation of the concept of matrix inversion to non-square matrices. In particular, when $L < D$ the matrix $A^T \cdot A$ is not invertible, but using the second limit in Eq. (4.11) when $\gamma \to 0^+$ we obtain

$$\boldsymbol{\theta}_2^{MAP} = \lim_{\gamma \to 0^+} A^T \cdot \left(A \cdot A^T + \gamma \mathcal{I}_L\right)^{-1} \boldsymbol{y} = A^T \cdot \left(A \cdot A^T\right)^{-1} \boldsymbol{y} \ , \tag{4.12}$$

where now the $L \times L$ matrix $A \cdot A^T$ is generically invertible; hence, when $L < D$, a particular solution of Eq. (4.7) is selected by the $L_2$ prior in the limit $\gamma \to 0^+$.

### 4.1.3  $L_1$-norm prior: lasso regression

In the $p = 1$ case, the $L_1$ prior is a product of exponential distributions over the absolute values of the parameters:

$$p(\boldsymbol{\theta}) = \prod_{d=1}^{D} \frac{\gamma}{2} e^{-\gamma |\theta_d|} \ . \tag{4.13}$$

Linear regression via MLE with a $L_1$ prior is also called lasso regression. As in the case of the $L_2$ prior, the log-posterior $\log P(\boldsymbol{y}|\boldsymbol{\theta}) + \log P(\boldsymbol{\theta})$ defined from Eqs. (4.5) and (4.13) is a concave function of $\boldsymbol{\theta}$, and has a unique maximum, but contrary to the $L_2$ case, no general analytical expression for the maximum exists.

Let us thus look at the simple case $L = D = 1$, with $A_{11} = 1$ to get some intuition about the effect of the prior. The MAP estimator is given by

$$\theta_1^{MAP}(y) = \underset{\theta}{\text{argmax}}\left[-\frac{1}{2}(y-\theta)^2 - \gamma|\theta|\right] = \begin{cases} y + \gamma & \text{if } y \leq -\gamma\,, \\ 0 & \text{if } |y| \leq \gamma\,, \\ y - \gamma & \text{if } y \geq \gamma\,. \end{cases} \qquad (4.14)$$

As a result of the presence of the $L_1$ prior, the inferred value of $\theta$ vanishes exactly over a finite-width range of $y$ values. This phenomenon is at odds with the effect of the $L_2$ prior, which gives, according to Eq. (4.10),

$$\theta_2^{MAP}(y) = \frac{y}{1+\gamma}\,. \qquad (4.15)$$

The two estimators are shown in figure 4.1. As expected they both reduce the value of the estimated parameter $\theta$ with respect to its maximum likelihood value, $\theta^{MLE} = y$ according to Eq. (4.8), but they do so in very different ways. $L_1$ sets $\theta_1^{MAP}$ to zero for data $y$ smaller (in absolute value) than $\gamma$, which acts as a threshold. For large $y$, its action is much less strong, as both $\theta^{MLE}$ and $\theta_1^{MAP}$ linearly grow with $y$ with unit slope. Conversely, the $L_2$ prior diminishes the slope for all values of $y$, without ever making $\theta_2^{MAP} = 0$, except for $y = 0$.

These behaviours can be understood from the shapes of the priors shown in figure 4.1. The tails of the priors at large $|\theta|$ decay much faster for $L_2$ than for $L_1$, hence disfavouring large values. Close to $\theta = 0$, the $L_2$ prior is very flat, which entails that a small, non-zero $\theta$ is very weakly disfavoured. The logarithm of the $L_1$ prior shows a marked cusp in $\theta = 0$, which is responsible for the existence of a finite-width range of values of $y$ such that $\theta_1^{MAP} = 0$, see figure 4.1. Quite generally, under the $L_1$ prior in Eq. (4.13) and for a generic likelihood function, the $d^{th}$ component of $\boldsymbol{\theta}_1^{MAP}$ vanishes if

$$-\gamma \leq \left.\frac{\partial}{\partial\theta_d}\log p(\boldsymbol{y}|\boldsymbol{\theta})\right|_{\boldsymbol{\theta}_1^{MAP}} \leq \gamma\,. \qquad (4.16)$$

Hence the $L_1$ prior favours explanation of data by zero-value parameters, and is useful for variable selection: it allows one to decide which among the $\theta_d$'s are relevant (non-zero) parameters. One practical way is to start with very high $\gamma$, *i.e.* with all parameters equal to zero. By progressively diminishing $\gamma$, more and more parameters start taking non-zero values. The procedure then halts when all relevant parameters have been "revealed". We will present one effective halting criterion in section 4.1.5.

### 4.1.4 Other values of $p$

$L_p$-norm priors can be considered for other values of $p$. On the one hand, all the values of $p \leq 1$ are qualitatively similar to $L_1$, see figure 4.1: they are characterised by a sharp cusp in $\theta = 0$, with a discontinuous derivative (diverging if $p < 1$). The case $p = 0$ is of particular interest, as $L_0$ precisely counts the number of non-zero components in $\boldsymbol{\theta}$, see Eq. (4.2). It is therefore a good idea to choose $p \leq 1$ to favour sparsity according

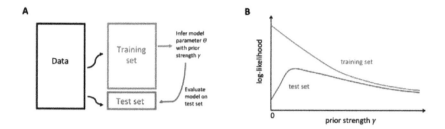

**Fig. 4.2** Principle of cross-validation. **A.** Data are partitioned into the training and test sets. The training set is used to infer $\boldsymbol{\theta}$ with prior strength $\gamma$. The inferred model can be then tested on the test set. **B.** Log-likelihood of the training and test sets as functions of the prior strength $\gamma$.

to the discussion above. On the other hand, many log-likelihood functions are concave functions of their parameters. This precious property ensures the existence and unicity of the global maximum, as well as its reachability by convex optimisation methods. It is thus natural to demand that the logarithm of the prior be also concave, which is the case for $p \geq 1$. The $L_1$ prior is thus the only one that favours sparsity while preserving concavity.

### 4.1.5 Determination of prior strength and cross-validation

An important issue is how to determine the adequate value of the parameters defining the prior, here, their strength $\gamma$. This can be done through a quite general procedure called cross-validation, see figure 4.2**A**. Cross-validation consists in dividing the data in two subsets: a large one, called training set, and a smaller one, called test set[2]. The model parameters are inferred from the training set only, while the quality of the inference is checked on the test set. Figure 4.2**B** shows the typical behaviour of the log-likelihood of the training and test sets as $\gamma$ varies. For low $\gamma$ values, the inferred model overfits the details of the training set and cannot generalise to other data, such as those in the test set: the log-likelihood of the training set is much higher than that of the test set. For large $\gamma$ the two log-likelihoods have similar low values: the model is overconstrained by the prior, and has not enough power to account for the true distribution of the data. In between, an optimal value for $\gamma$ exists, where the log-likelihood of the test set is maximal.

Typically, the training set contains 80 to 90% of the data, while the remaining data define the test set. This ratio is rather arbitrary but can neither be too small (model parameters would be poorly estimated, with a large variance of their posterior distribution) nor too large (the estimate of the log-likelihood of a small test set would be unreliable). The procedure can be repeated by partitioning the full data set in different ways, which allows one to estimate error bars on the log-likelihood curves in

---

[2]The dataset used to estimate the parameters that determine the learning process, also called hyperparameters, is sometimes called "validation set". The name "test set" is then reserved for a third set of data, that are never used during the learning, and are only used at the end to evaluate performance. The regularisation strength is an example of hyperparameter.

figure 4.2**B**. Notice that the partition must be done in such a way to make sure that the training and test sets are identical, statistically speaking. For instance, if the data are a series of time points, the training and test sets must be intertwined to ensure that, in any small time interval, say, 80% of data points are assigned to the training set, and 20% to the test set.

## 4.2 Conjugate priors

Likelihood functions are usually chosen based on two criteria: plausibility, that is, adequacy with the problem under consideration, and tractability. It is indeed crucial that calculations, either analytical or numerical, can be efficiently carried out, for instance, to maximise the likelihood or to sample it. A simple way to ensure that this property is not broken down by the introduction of the prior distribution is to make sure that the prior $P(\boldsymbol{\theta})$ has the same functional dependence on $\boldsymbol{\theta}$ as the likelihood $P(\boldsymbol{y}|\boldsymbol{\theta})$. Such priors are called conjugate priors. As we will see below, conjugate priors may be seen as a way to introduce additional data based on prior knowledge.

### 4.2.1 Binomial distributions

Consider, to start with, the case of the binomial distribution we have already studied in Chapter 1,

$$p(y|\theta) = \binom{M}{y} \theta^y (1-\theta)^{M-y} . \tag{4.17}$$

Here, $M$ is the total number of observations, and $y$ the number of positive events taking place with probability $\theta$. A conjugate prior should have the same functional dependence on $\theta$, and is generally of the form (up to a normalisation factor)

$$p(\theta) \propto \theta^{\alpha-1}(1-\theta)^{\beta-1} , \tag{4.18}$$

which corresponds to the Beta distribution we have already studied in Chapter 1. The parameters $\alpha$ and $\beta$ can take any positive value[3].

Multiplying the expressions in Eqs. (4.17) and (4.18) we find that the posterior distribution is functionally similar to the likelihood and can be written as

$$P(\theta|y) \propto \theta^{y'} (1-\theta)^{M'-y'} , \tag{4.19}$$

where the effective number of observations and of positive events are, respectively,

$$M' = M + \alpha + \beta - 2 \quad \text{and} \quad y' = y + \alpha - 1 . \tag{4.20}$$

In other words, the inclusion of the prior is equivalent to the addition of $\alpha + \beta - 2$ observations, out of which $\alpha - 1$ are positive events. This transformation of the data is also called a "pseudo-count", and we will see an application in tutorial 5.

---

[3] These formulae can be easily generalised to the case of multinomial distributions over $L = D$ variables, for which the likelihood reads

$$p(\boldsymbol{y}|\boldsymbol{\theta}) = \frac{M!}{\prod_{i=1}^{L} y_i!} \prod_i \theta_i^{y_i} \quad \text{with} \quad \sum_i y_i = M , \quad \sum_i \theta_i = 1 .$$

The conjugate prior is the Dirichelet distribution, $p(\boldsymbol{\theta}) = \frac{\Gamma(\sum_i \alpha_i)}{\prod_i \Gamma(\alpha_i)} \prod_{i=1}^{L} \theta_i^{\alpha_i - 1}$.

The value of $\theta$ that maximises the posterior is equal to $y'/M'$ according to the identities associated to the Beta distribution, see Eq. (1.25) and below. This result can be rewritten as

$$\theta^{MAP} = (1 - \kappa)\,\theta^{MLE} + \kappa\,\theta^{prior} ,\qquad (4.21)$$

where $\kappa = 1 - \frac{M}{M'}$ and

$$\theta^{MLE} = \frac{y}{M} , \qquad \theta^{prior} = \frac{\alpha - 1}{\alpha + \beta - 2} , \qquad (4.22)$$

are the estimators of the parameter obtained by maximising, respectively, the log-likelihood and the prior. The previous formula has a straightforward interpretation: the MAP estimator interpolates between the MLE estimator and the most likely $\theta$ fixed by the prior, with relative weights that depend on the ratio of the true and effective numbers of data. Such an interpolation, often referred to as "shrinkage", generally behaves better than the MLE estimator alone when the number of data is small, and this estimator is unreliable. Shrinkage is an empirical method that consists in transforming an estimator in the following way:

$$\text{estimator} \to (1 - \kappa) \times \text{estimator} + \kappa \times \text{assumption} . \qquad (4.23)$$

Here, $\kappa$ is a weighting coefficient that decreases from 1 to 0 as the number of data increases. This new estimator is pushed, or "shrunk" towards the assumption over $\theta$ coming from guesses and prior knowledge.

### 4.2.2 Exponential distributions

A large class of likelihood distributions is defined by the exponential family,

$$p(\boldsymbol{y}|\boldsymbol{\theta}) = e^{\boldsymbol{\theta}^T \cdot \boldsymbol{\Psi}(\boldsymbol{y}) - V(\boldsymbol{\theta})} \quad \text{with} \quad V(\boldsymbol{\theta}) \equiv \log \int d\boldsymbol{y} \, e^{\boldsymbol{\theta}^T \cdot \boldsymbol{\Psi}(\boldsymbol{y})} , \qquad (4.24)$$

where the integral can be replaced with a sum when the configurations $\boldsymbol{y}$ are discrete. The function $\boldsymbol{\Psi}(\boldsymbol{y})$ is called sufficient statistics: all configurations $\boldsymbol{y}$ with the same $\boldsymbol{\Psi}$ have the same probabilities. This is a very useful property when the dimension $D$ of $\boldsymbol{\Psi}$ is much smaller than the dimension $L$ of $\boldsymbol{y}$. For instance, the exponential family above contains the Gibbs distribution associated to the Ising model of statistical mechanics: $\boldsymbol{\Psi}(\boldsymbol{y})$ is then the set of single spins $y_i = \pm 1$ and pairs of spins $y_i y_j$, $\boldsymbol{\theta}$ the set of local fields $h_i$ and pairwise couplings $J_{ij}$, and $V(\boldsymbol{\theta})$ the logarithm of the partition function, see Eq. (2.20). If all the local fields and couplings have the same values, say, $h$ and $J$, as is the case in the so-called Curie-Weiss model, a sufficient statistic is the magnetisation $\boldsymbol{\Psi}(\boldsymbol{y}) = \frac{1}{L}\sum_i y_i$ of dimension $D = 1$, and $V$ can be calculated exactly. In the general case of non-equal fields and/or couplings, the dimension $D = \frac{1}{2}L(L+1)$ is very high and $V$ cannot be computed exactly, but one may resort to mean-field theory to define an approximation of $p$ with finite-dimensional sufficient statistics. The calculation of $V$ within mean-field theory can then be done in the thermodynamic limit $L \to \infty$.

Suppose we draw $M$ independent configurations $y^m$ from this exponential distribution. The maximum likelihood estimator for the model parameters is the root of the coupled set of implicit equations

$$F(\boldsymbol{\theta}^{MLE}) = \frac{1}{M} \sum_m \boldsymbol{\Psi}(y^m) , \quad \text{where} \quad F(\boldsymbol{\theta}) = \frac{\partial}{\partial \boldsymbol{\theta}} V(\boldsymbol{\theta}) \qquad (4.25)$$

is the gradient of $V$. Solving this equation may be quite hard from a computational point of view, as we will see in Chapter 5, but the solution is guaranteed to be unique due to the convexity of $V$ defined in Eq. (4.24)[4].

Let us now introduce the conjugate prior over the model parameters,

$$p(\boldsymbol{\theta}) \propto e^{\boldsymbol{\theta}^T \cdot \boldsymbol{\alpha} - \beta V(\boldsymbol{\theta})} . \qquad (4.26)$$

The maximum a posteriori estimator $\boldsymbol{\theta}^{MAP}$ of the model is the unique root of

$$F(\boldsymbol{\theta}^{MAP}) = \frac{1}{M + \beta} \left( \sum_m \boldsymbol{\Psi}(y^m) + \boldsymbol{\alpha} \right) = \frac{M}{M + \beta} F(\boldsymbol{\theta}^{MLE}) + \frac{\beta}{M + \beta} F(\boldsymbol{\theta}^{prior}) ,$$

$$(4.27)$$

where $\boldsymbol{\theta}^{prior}$ maximises the prior distribution. In the $D = 1$ case, according to the intermediate value theorem, there exists $\kappa \in [0; 1]$ such that Eq. (4.21) still holds. Notice, however, that $\kappa$ is not necessarily equal to $\beta/(M + \beta)$, but depends on the functional form of $V$.

### 4.2.3 Gaussian distribution

Consider the Gaussian distribution with parameters $\boldsymbol{\theta} = (\mu, \sigma^2)$, where $\mu$ is the mean and $\sigma^2$ the variance:

$$p(y|\boldsymbol{\theta}) = \frac{1}{\sqrt{2\pi\sigma^2}} e^{-\frac{(y-\mu)^2}{2\sigma^2}} . \qquad (4.28)$$

The dependence on $\mu$, at fixed $\sigma^2$, is Gaussian, and so will be the conjugate prior over $\mu$:

$$p(\mu) = \frac{1}{\sqrt{2\pi(\sigma^{prior})^2}} e^{-\frac{(\mu-\mu^{prior})^2}{2(\sigma^{prior})^2}} . \qquad (4.29)$$

The posterior $p(\mu|y)$ is obviously Gaussian, with mean

$$\mu^{MAP} = (1 - \kappa) \mu^{MLE} + \kappa \mu^{prior} , \qquad (4.30)$$

with $\mu^{MLE} = y$ according to Eq. (4.28), and $\kappa = \sigma^2/((\sigma^{prior})^2 + \sigma^2)$. This relationship extends to $M \geq 1$ data items $y^1, y^2, ..., y^M$. An easy calculation shows that the shrinkage-like Eq. (4.30) remains valid where $\mu^{MLE} = \sum_m y^m/M$ remains equal to the empirical mean, and $\kappa = \sigma^2/(M(\sigma^{prior})^2 + \sigma^2)$ decreases with $M$.

---

[4]It is easy to check that the Hessian matrix of $V$,

$$\frac{\partial^2}{\partial \theta_i \partial \theta_j} V(\boldsymbol{\theta}) = \langle (\Psi_i - \langle \Psi_i \rangle)(\Psi_j - \langle \Psi_j \rangle) \rangle \quad \text{with} \quad \langle \cdot \rangle = \int d\boldsymbol{y} \, (\cdot) \, e^{\boldsymbol{\theta}^T \cdot \boldsymbol{\Psi}(\boldsymbol{y})} \bigg/ \int d\boldsymbol{y} \, e^{\boldsymbol{\theta}^T \cdot \boldsymbol{\Psi}(\boldsymbol{y})} ,$$

is a covariance matrix and is therefore positive.

Let us now focus on the conjugate prior for the variance. It is convenient to change variable and consider the precision $t = 1/\sigma^2$ instead of the variance $\sigma^2$ itself. In fact, rewriting the likelihood in terms of the mean $\mu$ and the precision $t$, we have

$$p(y|\mu, t) = \sqrt{\frac{t}{2\pi}}\, e^{-\frac{t}{2}(y-\mu)^2} \ . \tag{4.31}$$

The conjugate prior for the precision, for a fixed mean, is therefore a Gamma distribution,

$$p(t) \propto t^{(\alpha-1)/2} e^{-\beta t/2} \ , \qquad \alpha, \beta > 0 \ , \tag{4.32}$$

which we have encountered in tutorial 1. The effect of this prior is to increase the effective number of data (by $\alpha - 1$) and the squared deviation to the mean (by $\beta$). The joint conjugate prior for the mean and precision is obtained, up to a normalisation factor, by multiplying Eqs. (4.29) and (4.32).

Finally, the above discussion can be extended to the case of multivariate Gaussian distributions associated to the precision matrix $T$, such as in Eq. (3.1). The likelihood of $M$ independently drawn configurations reads

$$p(Y|T) \propto (\det T)^{M/2}\, e^{-\frac{1}{2}\mathrm{Tr}(T \cdot Y^T \cdot Y)} \ , \tag{4.33}$$

where $Y$ is the $M \times L$ data matrix. The conjugate prior for the precision matrix is thus

$$p(T) = (\det T)^{(\alpha-1)/2}\, e^{-\frac{1}{2}\mathrm{Tr}(T \cdot \beta^{-1})} \ , \tag{4.34}$$

where $\beta^{-1}$ is a positive definite matrix. Comparison of Eqs. (4.33) and (4.34) shows that $\beta^{-1}$ can be interpreted as an added contribution to the empirical "covariance" matrix $\frac{1}{M} Y^T \cdot Y$, while $\alpha - 1$ plays the role, as in the $L = 1$ case above, of an additional number of data points. This prior is called the Wishart distribution, and is a generalisation of the Gamma distribution over positive real numbers to positive definite matrices.

## 4.3 Invariant priors

### 4.3.1 Change of priors under reparametrisation

Let us consider again the binomial distribution in Eq. (4.17), where $\theta$ is the probability of a positive event. As $\theta$ is comprised between 0 and 1, it may be natural in some circumstances to reparametrise the model. For instance, statistical physicists, who love spins and magnetic fields, would naturally replace $\theta$ by

$$h = \frac{1}{2} \log\left(\frac{\theta}{1-\theta}\right) \iff \theta = \frac{e^h}{e^{-h} + e^h} \ , \tag{4.35}$$

and the binomial distribution in Eq. (4.17) with

$$p(y|h) = \binom{M}{y} \frac{e^{h\,(2y-M)}}{(e^{-h} + e^h)^M} \ . \tag{4.36}$$

The field, or variable $h$, can now take any real positive or negative value, with $\theta = 0$ for $h = -\infty$ and $\theta = 1$ for $h = +\infty$.

Any prior distribution over $\theta$ induces a prior distribution over $h$ through the usual change-of-variable relation,

$$p(h) = p(\theta) \times \left| \frac{d\theta}{dh} \right| . \qquad (4.37)$$

This simple formula immediately tells us that choosing the uniform prior $p(\theta) = 1$, which seemed a rather innocuous assumption for $\theta$ in Chapter 1, induces a non-trivial distribution over $h$. An easy calculation shows that the associated prior over $h$ is

$$p(h) = \frac{1}{(e^{-h/2} + e^{h/2})^2} , \qquad (4.38)$$

which is plotted in figure 4.3. This distribution is far from being uniform, and does not look as natural as the uniform prior over $\theta$. Yet they describe exactly the same measure over the sets of $\theta$ values. Can we find a prior that is natural for all parametrisations?

### 4.3.2   Moving in the space of distributions

Consider the space of distributions $p(\boldsymbol{y})$ over the configurations $\boldsymbol{y}$. As $\theta$ varies along its $D$-dimensional manifold, a subspace is spanned by the likelihoods $p(\boldsymbol{y}|\boldsymbol{\theta})$, see figure 4.4. For simplicity, we focus in figure 4.4 and in the discussion below on the case $D = 1$.

Suppose we now sample $K$ model parameters $\theta^k$ from a prior $p(\theta)$ in the interval $[\theta_{min}; \theta_{max}]$. This sampling will result in the same number of distributions $p^k(\boldsymbol{y}) = p(\boldsymbol{y}|\theta^k)$, see figure 4.4. A uniform prior in the parameter space does not generally lead to a uniform sampling in the distribution subspace. This can be seen by estimating the dissimilarity between two successive distributions $p^k$ and $p^{k+1}$ through their Kullback-Leibler divergence. If $K$ is large, successive distributions are close to each other, and we may Taylor expand their divergence. The zeroth and first order terms in $\theta^{k+1} - \theta^k$ vanish due to the fact that $D_{KL}(p\|q)$ is minimal and null when $p = q$; we find to the lowest order in $\theta^{k+1} - \theta^k$ that

$$D_{KL}(p^k\|p^{k+1}) \simeq \frac{1}{2} I(\theta^k) \left( \theta^{k+1} - \theta^k \right)^2 , \qquad (4.39)$$

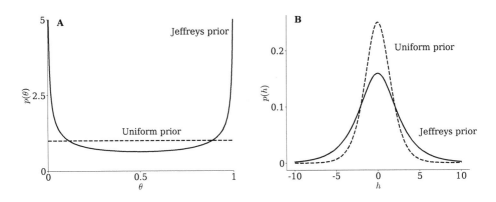

**Fig. 4.3  A**. Uniform and Jeffreys priors for the binomial distribution as function of $\theta \in [0; 1]$. **B**. Corresponding distributions upon the change of variable $\theta \to h$ of Eq. (4.35).

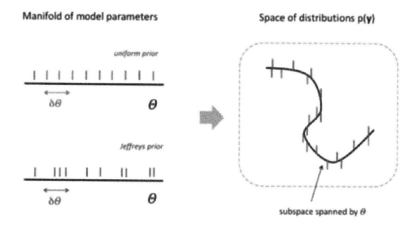

**Fig. 4.4** The space of distributions $p(\boldsymbol{y})$ and its subspace spanned by the likelihood distributions $p(\boldsymbol{y}|\boldsymbol{\theta})$ as $\boldsymbol{\theta}$ varies in some interval. As $\boldsymbol{\theta}$ is sampled from a prior distribution a sampling is induced in the space of distribution. A uniform prior in the parameter space generally corresponds to a non-uniform prior over the distribution space. Jeffreys prior, see Eq. (4.41), results in a uniform prior in the distribution space.

where $I(\theta^k)$ denotes the Fisher information associated to the likelihood $p(\boldsymbol{y}|\theta^k)$, see the definition in Eq. (2.33).

How does the spacing $\theta^{k+1} - \theta^k$ behave across the range $[\theta_{min}; \theta_{max}]$? In a small interval $\delta\theta$ the number of sampled points $\theta^k$ is proportional to $p(\theta) \times \delta\theta$. Equating this number with 1, we find that the spacing between two successive points is inversely proportional to the prior, $\theta^{k+1} - \theta^k \propto 1/p(\theta^k)$. We conclude that

$$D_{KL}(p^k || p^{k+1}) \propto \frac{I(\theta^k)}{p(\theta^k)^2} \ . \tag{4.40}$$

Uniform sampling in the $\theta$ space will result in undersampling in regions of the distribution space where the Fisher information is large, that is, where the likelihood $p(\boldsymbol{y}|\theta)$ strongly varies with $\theta$, and oversampling in low Fisher information regions. Reciprocally, we may compensate for these heterogeneities in the distribution space by tuning the prior distribution. According to Eq. (4.40) choosing the Jeffreys prior, defined through

$$p^{\text{Jeffreys}}(\theta) \propto \sqrt{I(\theta)} \ , \tag{4.41}$$

ensures that the sampling is now uniform in the distribution space. It is easy to check[5] that the Jeffreys prior behaves as expected from a probability distribution under a

[5] Differentiating twice the definition of the likelihood $p(y|h) \equiv p(y|\theta(h))$, we obtain

$$\frac{\partial^2}{\partial h^2} \log p(y|h) = \left(\frac{d\theta}{dh}\right)^2 \frac{\partial^2}{\partial\theta^2} \log p(y|\theta) + \left(\frac{d^2\theta}{dh^2}\right) \frac{\partial}{\partial\theta} \log p(y|\theta) \ .$$

The Fisher information for the $h$ variable is therefore

change of variable $\theta \to h$, that is,

$$\sqrt{I(h)} = \sqrt{I(\theta)} \times \left| \frac{d\theta}{dh} \right| . \tag{4.42}$$

As a final remark, while all the discussion above was limited to $D = 1$, it is clear that it can readily be extended to any dimension. The Jeffreys prior simply becomes

$$p^{\text{Jeffreys}}(\boldsymbol{\theta}) \propto \sqrt{\det I(\boldsymbol{\theta})} , \tag{4.43}$$

where $I(\boldsymbol{\theta})$ is now the $(D \times D)$-dimensional Fisher information matrix, see Eq. (2.45).

### 4.3.3   A couple of examples

As a first illustration let us calculate the Jeffreys prior associated to the binomial distribution of Eq. (4.17). Differentiating twice over $\theta$ and averaging over $y$ we get

$$I(\theta) = -\sum_y p(y|\theta) \frac{\partial^2}{\partial \theta^2} \log p(y|\theta) = \frac{M}{\theta(1-\theta)} . \tag{4.44}$$

Hence, the Jeffreys prior is given by, after normalisation,

$$p^{\text{Jeffreys}}(\theta) = \frac{1}{\pi \sqrt{\theta(1-\theta)}} , \tag{4.45}$$

and is plotted in figure 4.3. It diverges when $\theta \to 0$ or $\theta \to 1$, and is approximately flat in between. The graphical representation of the Jeffreys prior in the $h$ coordinate,

$$p^{\text{Jeffreys}}(h) = \frac{1}{\pi (e^{-h/2} + e^{h/2})} , \tag{4.46}$$

is also plotted in figure 4.3. It gives more weight to large $|h|$ compared to the distribution induced by the uniform prior over $\theta$.

Why is the Jeffreys prior giving more weight to $\theta$ close to 0 and 1? The standard deviation of a sample $y$ drawn from $p(y|\hat{\theta})$ is equal to $\sqrt{\hat{\theta}(1-\hat{\theta})}$, and it vanishes when $\hat{\theta}$ is close to the boundary values. Hence two distributions attached to two slightly different values $\hat{\theta}_1, \hat{\theta}_2$ in the vicinity of 0 or 1 generate more diverse samples than when they are close to, say, $\frac{1}{2}$. Again, we see that the Jeffreys prior favours regions in the parameter space corresponding to rapidly changing likelihood distributions.

As a second example, consider now the Gaussian distribution in Eq. (4.28). What is the Jeffreys prior for the mean $\mu$ (at fixed variance $\sigma^2$)? A quick calculation shows that

$$p^{\text{Jeffreys}}(\mu) = \text{constant} . \tag{4.47}$$

This shows that the uniform and Jeffreys prior may sometimes coincide. Notice that this prior is improper, as it cannot be normalised over real numbers.

$$I(h) = -\int dy \, p(y|h) \frac{\partial^2}{\partial h^2} \log p(y|h) = I(\theta) \left( \frac{d\theta}{dh} \right)^2 - \left( \frac{d^2\theta}{dh^2} \right) \int dy \frac{\partial}{\partial \theta} p(y|\theta) .$$

As the second term vanishes due to the normalisation of $p$, we end up with Eq. (4.42).

## 4.4  Tutorial 4: sparse estimation techniques for RNA alternative splicing

In eukaryotes, genes are composed of exon and intron sequences, as illustrated in figure 4.5. In the course of transcription to messenger RNA (mRNA), through a process called alternative splicing, introns are removed and some of the exons are retained. The set of retained exons can vary, resulting in the expression of different transcripts and different protein products from the same gene. Through RNAseq, a technology consisting in reverse-transcribing mRNA to DNA and sequencing the latter, one can get information about which transcripts are present in the cell. However, as sequenced reads are typically shorter than mRNA this technique does not allow one to fully access the transcripts, but only to estimate the numbers of single exons and exon-exon junctions (sequence reads that overlap two contiguous exons). The question we address in this tutorial is how to deduce the most probable set of transcripts compatible with these partial data [23]. We will use sparse matrix estimation techniques based on a $L_1$ prior (lasso regression) to do so.

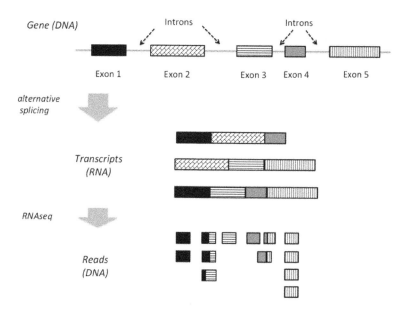

**Fig. 4.5** Illustration of alternative splicing in eukaryotic cells. Many different transcripts can be obtained by keeping or removing exons. Through RNAseq short reads can be obtained, giving information about the presence of single exons or exon-exon junctions. The number of copies varies from read to read.

### 4.4.1  Problem

We are going to discuss here a very simplified model of the transcription and sequencing process. Consider a gene containing $L$ exons. A transcript can be described by a list of $L$ binary digits, indicating if the exon is present or not in the transcript. As an example, for $L = 3$, the string 100 indicates the transcript that includes the first exon only. Note that the string of all zeros is not a possible transcript because there must be at least one exon in the transcript, hence there are in total $D = 2^L - 1$ possible transcripts. We assign to each transcript an index $t = 1, \ldots, D$ that is the natural number corresponding to the binary number encoded by the transcript. Notice that $t$ reads from left to right, *e.g.* for $L = 3$, $t = 1$ corresponds to 100, $t = 2$ corresponds to 010, $t = 3$ corresponds to 110, and so on.

**Case of non-overlapping reads** – We first assume that each read of DNA corresponds to a single exon. In other words, each exon present in the cell, irrespectively of the transcript it belongs to, is read separately in the sequencing. The experiment then gives access to numbers $y_i$, $i = 1, \cdots, L$, which give the total number of copies of exon $i$ that are present in the cell. We define the $L \times D$ matrix $A$ of all possible transcripts, where $A_{i,t} = 1$ if the exon in position $i = 1, \ldots, L$ is present in the transcript $t = 1, \ldots, 2^L - 1$, and $A_{i,t} = 0$ otherwise. In other words, the $t^{th}$ column of the matrix $A$ corresponds to the $L$-bit-string representation of the transcript $t$. We denote by $\boldsymbol{\theta} = \{\theta_t\}$ the vector of the (unknown) numbers of copies of transcripts $t$ in the cell (assumed to be proportional to the number of proteins of a particular type). Because transcripts of type $t$ contribute $A_{i,t} \times \theta_t$ exons of type $i$, we have

$$y_i = \sum_{t=1}^{2^L - 1} A_{i,t} \theta_t , \qquad \text{or} \qquad \boldsymbol{y} = A \boldsymbol{\theta} . \tag{4.48}$$

Here $\boldsymbol{y}$ is measured by sequencing, $A$ is known and $\boldsymbol{\theta}$ is the unknown vector we want to infer. The entries $\theta_t$ are zero if transcript $t$ is not present in the cell, and give the transcript copy number otherwise. We do not know the possible set of transcripts a *priori*, but we know that such a set is very small (typically of the order of 10) with respect to the total number of possible transcripts, which is $D = 2^L - 1$. Hence, $\boldsymbol{\theta}$ is a sparse vector with many zero entries. Given $\boldsymbol{y}$, and assuming that the measurement is corrupted by noise, the problem of reconstructing $\boldsymbol{\theta}$ falls in the class of linear regression discussed in section 4.1.1.

The sparsity of the vector $\boldsymbol{\theta}$ can then be enforced through a $L_1$-norm regularisation. Together with Eq. (4.48), this leads to the minimisation of the quadratic form

$$\min_{\boldsymbol{\theta}} \left\{ \frac{1}{2L} \sum_i \left( y_i - \sum_t A_{i,t} \theta_t \right)^2 + \gamma \sum_t |\theta_t| \right\} , \tag{4.49}$$

which we will perform using the lasso regression algorithm discussed in section 4.1.3. By developing the square, Eq. (4.49) is equivalent to

$$\min_{\boldsymbol{\theta}} \left\{ \frac{1}{2} \sum_{t,t'} X_{t,t'} \theta_t \theta_{t'} - \sum_t z_t \theta_t + \gamma \sum_t |\theta_t| \right\} , \tag{4.50}$$

where $X_{t,t'} = \sum_i A_{i,t} A_{i,t'} / L$ and $z_t = \sum_i y_i A_{i,t} / L$. The quadratic form in Eq. (4.49) is convex and therefore it has a unique minimum; one can try to find it by solving

Eq. (4.49) iteratively for each $t$, a method known as *coordinate descent, i.e.* iterative minimisation along each coordinate $\theta_t$. This minimisation can be done analytically. Differentiating the quadratic form in Eq. (4.50) with respect to $\theta_t$ we have

$$X_{t,t}\,\theta_t + \sum_{t'(\neq t)} X_{t,t'}\,\theta_{t'} - z_t + \gamma\,\text{sign}(\theta_t) = 0\ , \tag{4.51}$$

which is solved by

$$\theta_t = \begin{cases} \frac{v_t - \gamma\,\text{sign}(v_t)}{X_{t,t}} & \text{for } |v_t| > \gamma, \\ 0 & \text{otherwise}, \end{cases} \tag{4.52}$$

where $v_t = z_t - \sum_{t'(\neq t)} X_{t,t'}\theta_{t'}$.

**Case of overlapping reads** – We now take into account the presence of DNA reads overlapping two exons. In addition to the $L$ single-exon reads, there are $L(L-1)/2$ such exon-exon junction reads, giving $L_t = L(L+1)/2$ different transcript types. Of course, the total number of possible transcripts is unchanged and remains equal to $D = 2^L - 1$.

A redundant but convenient representation is obtained, by analogy with the single exon case described above, by writing a transcript as a binary number with $L_p$ digits. The first $i = 1, \cdots, L$ binary entries of the $t^{th}$ column of the matrix $A_{i,t}$, corresponding to transcript $t$, indicate which exons are present, and the remaining $i = L+1, \cdots, L(L+1)/2$ entries indicate which junctions between adjacent exons are present. As an example, for $L = 3$ exons, we represent the $D = 7$ possible transcripts as strings of $L_p = 6$ bits, leading to a matrix $A_{i,t}$ of dimension $L_p \times D$ with the following columns (here, shown horizontally for typesetting reasons):

$$\begin{array}{ll} 100000 & \text{exon 1 is present,} \\ 010000 & \text{exon 2 is present,} \\ 001000 & \text{exon 3 is present,} \\ 110100 & \text{exons 1 and 2 are present,} \\ 101010 & \text{exons 1 and 3 are present,} \\ 011001 & \text{exons 2 and 3 are present,} \\ 111101 & \text{all exons are present.} \end{array} \tag{4.53}$$

The first three bits are identical to the non-overlapping case, while the last three encode the information about overlaps. With this choice, exactly as in the previous case, transcripts of type $t$ contribute $A_{i,t} \times \theta_t$ reads of type $i$. In the example above, a number $\theta_t$ of transcripts with exons 1 and 2 contribute $\theta_t$ reads of exon 1 ($i = 1$), $\theta_t$ reads of exon 2 ($i = 2$) and $\theta_t$ reads of a junction between exons 1 and 2 ($i = 4$). Notice that, in reality, the number of reads of a given type produced per transcript are likely to depend on the length of the exons, and will therefore vary with the read type. We neglect this effect here for the sake of simplicity.

The vector $y_i$ encoding the number of recorded exons and exon junctions, $i = 1, \ldots, L_p$, is given as before by $\boldsymbol{y} = A\boldsymbol{\theta}$ and we can apply the lasso algorithm to find the sparse vector $\boldsymbol{\theta}$. The additional information about exon junctions helps reducing the degeneracy of the problem. As an illustration, consider $L = 3$, and suppose that we measure $y_1 = y_2 = 5$ and we know that there are no transcripts with all the three exons. Without information on junctions, this result could come either from 5 transcripts with exon 1 only and 5 transcripts with exon 2 only, or alternatively from 5 transcripts with both exon 1 and exon 2, or any combination of these two extreme cases. If we measure $y_4$, we can fix the number of transcripts with both exon 1 and exon 2, and deduce the number of transcripts with exon 1 only $(y_1 - y_4)$ or exon 2 only $(y_2 - y_4)$.

**Data:**
Data files can be downloaded from the book webpage[6], see tutorial 4 repository. We constructed simulated data for a gene composed by $L = 10$ exons; we chose uniformly at random $N_t$ transcripts, each containing between $N_e^{\min}$ and $N_e^{\max}$ exons, hence with $N_e = (N_e^{\min} + N_e^{\max})/2$ exons on average. We generated the corresponding true $\widehat{\theta}_t$ at random, and constructed $\boldsymbol{y} = A\widehat{\boldsymbol{\theta}}$.

- Data without overlaps: the data file *dataNe2Nt5-nooverlap.dat* contains the measured values of $y_i$, for $i = 1, \ldots, L = 10$.
- Data with overlaps: the data file *dataNe2Nt5.dat* contains the measured values of $y_i$, with $i = 1, \ldots, L_p$ and $L_p = L(L+1)/2 = 55$.
- The original set of transcripts $t$ and their corresponding expressions $\theta_t$ are given in the file *transcriptscpnbNe2Nt5.dat*. The format of the file is the following. The first column is the index $t$ of the pattern, represented as a number $t = 0, \ldots, 1022 = 2^L - 2$ (note that because we use the Python language, in the numerical calculations the index $t$ is shifted by one with respect to the convention used in the theoretical calculations, in such a way that $t = 0$ corresponds to the first transcript, and so on). The second column is the corresponding expression level $\widehat{\theta}_t$.

**Questions:**
In the tutorial 4 data repository, the Jupyter notebook *tut4_start.ipynb* reads the data and constructs the matrix $A$; it can be used to start working more quickly. Complete the notebook or write your own code to answer the following questions.

1. Program the lasso routine following the procedure described above. Start with $\theta_t = 0$ and then update the $\theta_t$ sequentially according to Eq. (4.52), until convergence. Compare the results with the lasso routine given in the *Scikit-learn* Python library, *linear_model.Lasso*, for some simple choice of the matrix $A$ and the vector $\boldsymbol{\theta}$.
*Optional:* Note that one can include in the $L_1$ regularisation prior the additional information that $\theta_t \geq 0$, which is relevant for this application. Discuss the changes that can be made in this case, to further simplify the procedure.

2. Consider the case of non-overlapping reads, by reading the corresponding data files. Write the matrix $A$ and implement the lasso algorithm using either your routine or

---

[6]https://github.com/StatPhys2DataDrivenModel/DDM_Book_Tutorials

the *Scikit-learn* one. Discuss the solutions for $\boldsymbol{\theta}$ given by the lasso algorithm for some values of $\gamma = 0.01, 0.1, 1$, and in what they differ from the original vector $\widehat{\boldsymbol{\theta}}$ used to generate the data. In particular:

- Compare $\boldsymbol{\theta}(\gamma)$ with the true $\widehat{\boldsymbol{\theta}}$.
- Rank the $\theta_t(\gamma)$ from the largest to the smallest. Compute the number of truly predicted transcripts in the top $2N_t$ predictions.
- *Optional:* Make a plot of the "positive predictive values": starting from the ordered list of $\theta_t$, check sequentially if $\theta_t$ is a true prediction (or equivalently one of the original transcripts) or not, and plot as a function of the prediction rank the percentage of truly predicted transcripts.

3. Rewrite the minimisation of the quadratic form in Eq. (4.49) using the Bayes' rule, identify the $L_1$ penalty as a "prior" probability and discuss the consequences of such penalty on the solution $\boldsymbol{\theta}$ and the role of the regularisation parameter $\gamma$. In particular compare the $L_1$ penalty with a $L_2$ penalty and discuss why the $L_1$ penalty is better than the $L_2$ penalty for this problem.

4. Discuss more systematically the dependence on $\gamma$, by doing the following.

- Plot the mean square error (MSE) as a function of $\gamma$:

$$\mathrm{MSE}(\gamma) = \frac{1}{2^L - 1} \sum_{t=1}^{2^L - 1} [\widehat{\theta}_t - \theta_t(\gamma)]^2 , \tag{4.54}$$

and find the value $\gamma^*$ that minimises the MSE.
- Plot the quadratic cost function $C(\gamma)$ of the solution $\boldsymbol{\theta}$ as a function of the regularisation strength $\gamma$, with

$$C(\gamma) = \sum_i \frac{1}{2L} \left[ y_i - \sum_t A_{i,t}\theta_t(\gamma) \right]^2 + \gamma \sum_t \theta_t(\gamma) , \tag{4.55}$$

and also discuss the two terms of $C(\gamma)$ separately by making a parametric plot of the solution norm (second term of the equation above) as a function of the first term, also called residual norm. Discuss the values of the two terms for the optimal value of $\gamma$.
- Plot the total number of non-zero elements of $\theta_t$ as a function of $\gamma$, and plot the value of $\theta_t$ as a function of $\gamma$ for the 5 transcripts corresponding to the true transcripts ($t = 7, 16, 18, 127, 267$).

5. Read the full data with overlapping reads. Write the matrix $A$ in this case and solve the lasso problem to find the vector $\boldsymbol{\theta}$. Repeat the same study as in points 2, 3 and 4 above.

6. *Optional:* Add some noise to the vector $\boldsymbol{y}$ and study the robustness of the results with respect to the noise.

## 4.4.2    Solution

The results of the lasso algorithm depend on the convergence criterion, especially for small $\gamma$ and in the non-overlapping case, where one can have multiple almost equivalent solutions, leading to problems in convergence for the lasso routine. In the following, we used the *Scikit-learn* routine with default parameters except *max_iter*= 100000, which provides a good compromise between fast and proper convergence. The reader can discuss, an as optional exercise, how the results depend on the maximum number of iterations or on the convergence threshold.

### Question 1: lasso routine
Our implementation of the lasso routine can be found in the *tutorial4.ipynb* notebook, where we also compare it with the *Scikit-learn* implementation finding identical results. The condition $\theta_t \geq 0$ in the prior is simply implemented by modifying Eq. (4.52) as

$$\theta_t = \begin{cases} \frac{v_t - \gamma}{X_{t,t}} & \text{for } v_t > \gamma, \\ 0 & \text{otherwise.} \end{cases} \tag{4.56}$$

### Question 2: Results for several values of the regularisation strength
We first report results in the case of no overlap, with $\gamma = 0.01, 0.1, 1$, in figure 4.6. The true transcripts are labelled by the indices $\{7, 16, 18, 127, 267\}$ with expressions, respectively, $\{14, 327, 344, 275, 26\}$. Therefore, 16, 18 and 127 are the mostly expressed transcripts and should be easier to identify, while 7 and 267 have low expressions and are probably harder to identify.

The 10 transcripts with the largest components $\theta_t(\gamma)$ are, respectively:

- $\gamma = 0.01$: $\{16, 143, 2, 18, 144, 17, 130, 146, 402, 10\}$.
- $\gamma = 0.1$: $\{18, 16, 144, 146, 402, 26, 22, 154, 410, 30\}$.
- $\gamma = 1$: $\{18, 16, 144, 146, 402, 26, 22, 154, 410, 30\}$.

We observe that transcripts 16 and 18 are always correctly identified, and with higher regularisation they correspond to the two top predictions. The results for $\gamma = 0.1$ and $\gamma = 1$ are very similar.

### Question 3: Bayesian interpretation of the lasso algorithm
In the Bayesian inference framework, the $L_1$-norm regularisation, which we have added to the quadratic cost function to impose the sparsity of the vector $\boldsymbol{\theta}$, can be considered as a prior. We can rewrite Eq. (4.49) by using Bayes' rule as:

$$p(\boldsymbol{\theta}|\boldsymbol{y}) = \frac{p(\boldsymbol{y}|\boldsymbol{\theta}) \times p(\boldsymbol{\theta})}{p(\boldsymbol{y})} \tag{4.57}$$

with

$$p(\boldsymbol{y}|\boldsymbol{\theta}) \propto \exp\left\{-\frac{1}{2L}\sum_i \left(y_i - \sum_t A_{i,t}\theta_t\right)^2\right\}, \tag{4.58}$$

which implicitly assumes that $y_i = \sum_t A_{i,t}\theta_t + \xi_i$, where $\xi_i$ is a Gaussian noise of variance $L$, and the prior probability

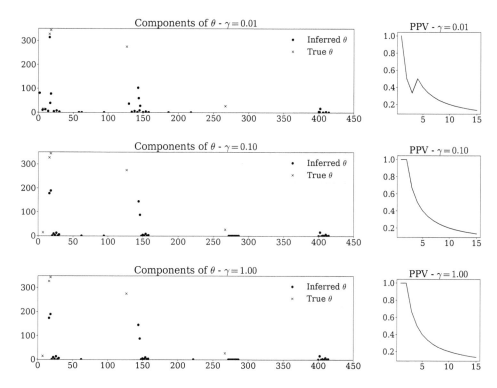

**Fig. 4.6** Data with non-overlapping reads. A comparison of the true $\hat{\theta}$ and inferred $\theta$ (the vector components are plotted against their number) is shown in **A** ($\gamma = 0.01$), **C** ($\gamma = 0.1$), **E** ($\gamma = 1$). PPV curves, which give the fraction of correct predictions (*i.e.* non-zero components $\hat{\theta}_t$) among the largest $x$ components of $\theta_t$ ($x$ is the horizontal axis), are shown in **B** ($\gamma = 0.01$), **D** ($\gamma = 0.1$), **F** ($\gamma = 1$).

$$p(\boldsymbol{\theta}) \propto \exp\left\{-\gamma \sum_t \theta_t\right\}, \qquad \text{for } \theta_t \geq 0. \tag{4.59}$$

Such prior is called $L_1$ prior, and it is more generally written with the absolute value of $\theta_t$, but for the problem considered here we can impose that $\theta_t$ is positive, because it corresponds to the number of transcripts of type $t$. Maximising the posterior then corresponds to minimising the quadratic form in Eq. (4.49). The $L_1$ prior has two consequences on the solution $\boldsymbol{\theta}$: it imposes a sparse solution by adding a penalty that increases quickly near the origin, with a linear slope. This is different from the quadratic penalty called $L_2$-norm, which increases only quadratically near the origin. Moreover the $L_1$ prior penalises large values of $\theta_t$, because the prior is an exponentially decaying function with a characteristic length of $1/\gamma$, which fixes the maximal ranges of the values of $\theta_t$ to the order of $1/\gamma$. On the contrary, the $L_2$-norm grows faster and thus penalises more strongly the large values of $\theta_t$, hence typical value of $\theta_t$ will be of the order of $1/\sqrt{\gamma}$. Note that using the condition $\theta_t \geq 0$ we can simplify Eq. (4.52) as

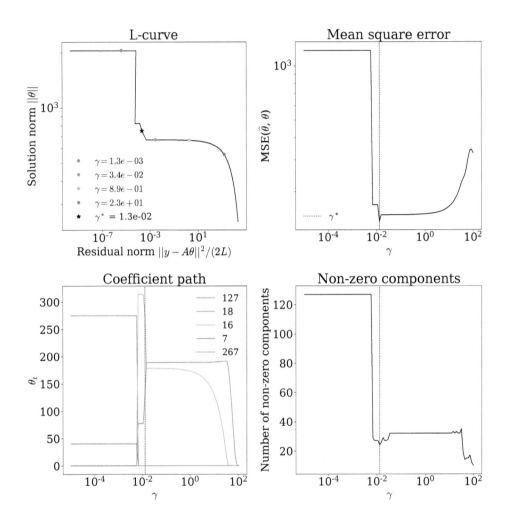

**Fig. 4.7** Data with non-overlapping reads. **A.** Parametric curve, obtained by varying the regularisation parameter $\gamma$, of the solution norm versus the residual norm. **B.** Mean-square error of the predicted transcript (or protein) copy numbers with respect to the true ones. **C.** Variation of the copy number of the transcripts (proteins) as a function of the regularisation parameter $\gamma$ for the 5 transcripts that are really present in the data $(t = 7, \theta_t = 14; t = 16, \theta_t = 327; t = 18, \theta_t = 344; t = 127, \theta_t = 275; t = 267, \theta_t = 26)$. **D.** The number of non-zero inferred copy numbers as a function of $\gamma$.

discussed in the solution of question 1 above.

## Question 4: Dependence of the optimal solution on regularisation strength

Results for non-overlapping reads are given in figures 4.7. The optimal value of $\gamma^\star = 1.3 \cdot 10^{-2}$, which minimises the mean square error in Eq. (4.54), corresponds to a typical decay range $\theta_t^\star \sim 100$, which is of the order of the typical values of the number of transcripts in the real data. Indeed, in this case of non-overlapping reads, imposing a sparse solution allows us to remove the degeneracy by favouring transcripts where multiple exons are present, which was indeed the case in the data. As can be seen in figure 4.7, upon increasing the parameter $\gamma$ the solution becomes sparser and sparser. Note that very small $\theta_t < 10^{-14}$ have been rounded to zero. The L-curve (parametric plot of the solution norm versus the residual norm) shows that the optimal value of $\gamma$ realises a compromise between the minimisation of the residual norm (difference between the predicted reads and the measured reads) and the minimisation of the norm that is imposed by the prior.

## Question 5: Overlapping case

Having additional information on which exons are nearby in the transcript allows one to remove the degeneracy of possible solutions, which is present in the non-overlapping case. In the case of overlapping reads, the solution is thus less degenerate and the two lasso routines (our implementation and the *Scikit-learn* package) converge more easily and to the same (or almost the same) solution. As shown in figure 4.8, the predictions largely improve: the top 5 predicted transcripts corresponds to true ones for $\gamma = 0.01$, and the performance is slightly degraded upon increasing $\gamma$. In this case, we have so much information that adding a too strong prior results in a worse prediction. Indeed, figure 4.9 shows that already for very small values of $\gamma$ one obtains a good solution.

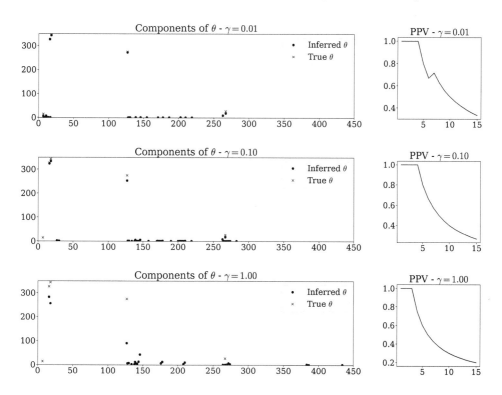

**Fig. 4.8** Data with overlapping reads, same as figure 4.6.

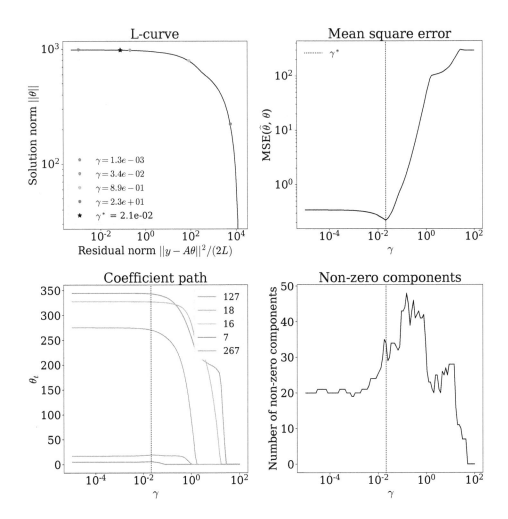

**Fig. 4.9** Data with overlapping reads, same as figure 4.7.

The optimal value $\gamma^\star = 2.1 \cdot 10^{-2}$ is related, also in this case, to the order of magnitude of the largest component in the data, $\theta_t^\star \sim 100$. When adding a noise (optional) in the data, the optimum on $\gamma$ become once again more pronounced.

# 5

# Graphical models: from network reconstruction to Boltzmann machines

Understanding how the many elementary components of a system, be they neurons, genes, species, etc., interact to produce a global behaviour is the central scope of data-driven modelling. A simple way to characterise these interactions is to look at the correlations between pairs of components. Yet, pairwise correlations are potentially misleading, as they can reflect indirect effects mediated via third-body components, rather than direct interactions. A more sensible estimate of interactions is provided by the graph of dependencies we have introduced in the context of multivariate Gaussian distributions in Chapter 3. Briefly speaking, the idea is to look at the conditional probabilities of a component, or variable, say, $y_i$, given the other variables $y_j$ with $j \neq i$. The graph of dependencies characterises how much $y_j$ conditions the distribution of $y_i$, see figure 3.2. The question we will ask below is: how can this graph, or network, be reconstructed from a set of observations of the variables?

## 5.1 Network reconstruction for multivariate Gaussian variables

As in Chapter 3, we consider a set of random variables $\boldsymbol{y} = (y_1, \cdots, y_L)$ drawn from the centred Gaussian multivariate distribution given by Eq. (3.1). This distribution is fully defined by the symmetric precision matrix, $\widehat{T}$, of dimension $L \times L$. Suppose we now generate $M$ i.i.d. data points, which we stack in the data matrix $Y = \{\boldsymbol{y}^m\}_{m=1,\cdots,M}$, of dimension $M \times L$.

Our goal is to infer the precision matrix from the data, in particular its off-diagonal elements $\widehat{T}_{ij}$ that enter the conditional probabilities of single variables,

$$
p(y_i | \boldsymbol{y}_{\setminus i}; \widehat{T}) = \sqrt{\frac{\widehat{T}_{ii}}{2\pi}} \exp\left[ -\frac{\widehat{T}_{ii}}{2} \left( y_i + \sum_{j(\neq i)} \frac{\widehat{T}_{ij}}{\widehat{T}_{ii}} y_j \right)^2 \right], \tag{5.1}
$$

where $\boldsymbol{y}_{\setminus i} = (y_1, \cdots, y_{i-1}, y_{i+1}, \cdots, y_L)$ is the set of $L-1$ variables excluding $y_i$, and define the pairwise dependencies.

### 5.1.1 Maximum likelihood estimator

The log-likelihood for a candidate precision matrix $T$, normalised per data point, is

$$\mathcal{L} \equiv \frac{1}{M} \log p(Y|T) = \frac{1}{M} \sum_{m=1}^{M} \log p(\boldsymbol{y}^m|T) \ . \tag{5.2}$$

$\mathcal{L}$ depends on the data $Y$ through the empirical covariance matrix $C = \frac{1}{M} Y^T \cdot Y$ in Eq. (3.17), and is equal to

$$\mathcal{L}(C,T) = \frac{1}{2} \log \det T - \frac{1}{2} \mathrm{Tr} \left( T \cdot C \right) \ , \tag{5.3}$$

up to an irrelevant additive term. Let us compute the maximum likelihood estimator (MLE) for $T$ by maximising the likelihood $\mathcal{L}(C,T)$.

First, we want to show that $\mathcal{L}$ is a strictly concave function of $T$, and has therefore a unique maximum. Consider the restriction $\ell$ of $\mathcal{L}$ to the segment joining two distinct symmetric positive matrices $T_1$ and $T_2$,

$$\ell(\alpha) \equiv \mathcal{L}(\alpha T_1 + (1-\alpha)T_2, C) \ , \qquad \alpha \in [0,1] \ . \tag{5.4}$$

Denoting by $\{\lambda_i\}_{i=1,\cdots,L}$ the eigenvalues of $T_2^{-1} \cdot T_1$, we have

$$\begin{aligned} \ell''(\alpha) &= \frac{1}{2} \frac{d^2}{d\alpha^2} \left( \log \det[\alpha T_1 + (1-\alpha)T_2] - \mathrm{Tr} \left[ (\alpha T_1 + (1-\alpha)T_2) \cdot C \right] \right) \\ &= \frac{1}{2} \frac{d^2}{d\alpha^2} \left( \log \det T_2 + \log \det[\mathcal{I}_L + \alpha(T_2^{-1} \cdot T_1 - 1)] \right) \\ &= \frac{1}{2} \frac{d^2}{d\alpha^2} \sum_{i=1}^{L} \log \left( 1 + \alpha(\lambda_i - 1) \right) = -\frac{1}{2} \sum_{i=1}^{L} \frac{(\lambda_i - 1)^2}{\left( 1 + \alpha(\lambda_i - 1) \right)^2} \ . \end{aligned} \tag{5.5}$$

As $T_1 \neq T_2$, the product of matrices $T_2^{-1} \cdot T_1$ is not identical to the identity $\mathcal{I}_L$, and at least one of the eigenvalues $\lambda_i$ is different from one. Hence, the second derivative of $\ell(\alpha)$ is strictly negative, and $\ell(\alpha)$ is strictly concave. As $T_1$ and $T_2$ can be chosen arbitrarily in the space of positive definite matrices, the log-likelihood $\mathcal{L}(T,C)$ is strictly concave.

We may now determine the unique maximum of $\mathcal{L}$. Using the general identity[1]

$$\frac{\partial}{\partial T_{ij}} \log \det T = (T^{-1})_{ji} \ , \tag{5.6}$$

and the fact that the matrix $T$ is symmetric[2] we see that $T^{MLE}$ is the root of

$$\frac{\partial \mathcal{L}}{\partial T_{ij}} (T^{MLE}, C) = 0 = C_{ji} - (T^{-1})_{ji} \ , \tag{5.7}$$

---

[1] The proof is obtained by considering, for a matrix $\Delta$ with small $L_2$ norm,

$$\log \det(T + \Delta) - \log \det T = \log \det(\mathcal{I}_L + \Delta \cdot T^{-1}) = \mathrm{Tr} \left( \Delta \cdot T^{-1} \right) + O(|\Delta|^2) \ .$$

Differentiating both sides of this equation with respect to $\Delta_{ij}$ and setting $\Delta = 0$ gives Eq. (5.6).

[2] Because $T$ is symmetric, we consider that its entries in the upper triangle $i < j$ are independent, and mirrored in the lower triangle $i > j$. Differentiating with respect to $T_{ij}$ amounts therefore to differentiate with respect to both $T_{ij}$ and $T_{ji}$, hence the factor 2 between Eq. (5.6) valid for a generic matrix and Eq. (5.7) obtained for a symmetric precision matrix.

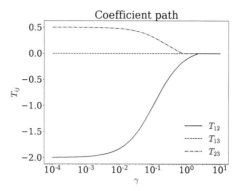

**Fig. 5.1** Evolution of the off-diagonal matrix elements of the precision matrix $T$ maximising Eq. (5.9) with $\gamma$, for the $3 \times 3$ covariance matrix $C$ corresponding to the multivariate Gaussian distribution of figure 3.2. The first non-zero interaction, $T_{12}$, is uncovered when $\gamma < \gamma_1 = \frac{20}{9} \simeq 2.22$. The support of the dependency graph is unveiled as soon as $\gamma < \gamma_2 \simeq 0.777$, compare with the left panel of figure 3.2. However, the true interactions coefficient are found back for $\gamma = 0$ only.

for all $i, j$. Therefore, the MLE estimator is the inverse of the empirical correlation matrix,

$$T^{MLE} = C^{-1} . \tag{5.8}$$

If $C$ has zero eigenvalues, *e.g.* if $M < L$, $T$ is ill-defined, and regularisation is needed, a situation we have already encountered for linear regression in Chapter 4. Various priors can be considered, such as the Wishart distribution over definite positive matrices in Eq. (4.34), which is conjugate to the likelihood. We hereafter consider the case of the $L_1$-norm based prior, which is of particular interest if one looks for sparse approximations of $T$.

## 5.1.2 $L_1$-norm regularisation: graphical lasso

As discussed in Chapter 4, it is a wise idea to add a regularisation in order to avoid overfitting and, if desired, enforce sparsity of the interaction matrix. With the $L_1$-norm regularisation, the log-posterior for $T$ reads

$$\mathcal{L}(T, C, \gamma) = \frac{1}{2} \log \det T - \frac{1}{2} \mathrm{Tr}\,(C \cdot T) - \gamma \sum_{i<j} |T_{ij}| . \tag{5.9}$$

The regularisation has been added to the $\frac{1}{2}L(L-1)$ off-diagonal parameters of $T$ only, since diagonal parameters have no reason to be sparse. $\mathcal{L}$ is a concave function, its maximum is therefore unique and can be efficiently found with numerical procedures. One of these algorithms is called graphical lasso or g-lasso [24].

Let us discuss the role of the regularisation penalty $\gamma$, which is illustrated in figure 5.1 for $L = 3$ variables. For very large $\gamma$, the maximum of $\mathcal{L}$ is located in $T_{diag}$, the matrix whose off-diagonal elements vanish and diagonal elements are given by

$T_{ii} = 1/C_{ii}$. According to Eq. (5.7) the gradient of the log-likelihood $\mathcal{L}$ with respect to $T_{ij}$ is equal to $C_{ij}$ for $i \neq j$. We deduce from Eq. (4.16) that, upon decreasing $\gamma$, a first non-zero off-diagonal entry of $T$ appears when $\gamma$ reaches the critical value

$$\gamma_1 = \max_{i<j} \left| \frac{\partial L}{\partial T_{ij}} (T_{diag}, C, \gamma) \right| = \max_{i<j} |C_{ij}| . \tag{5.10}$$

Hence, the first link appearing in the inferred dependency network connects the two most correlated pair of variables. Upon further reducing $\gamma$, this non-zero element grows (in absolute value), while the other coefficients of $T$ remain equal to zero. At a second transition point $\gamma_2 < \gamma_1$ a second off-diagonal entry of $T$ emerges. These two non-zero elements then evolve, and so on, until all non-zero entries of $T$ are revealed. These trajectories can be efficiently computed numerically to obtain the whole g-lasso regularisation path.

At which value of $\gamma$ should the inference procedure be stopped? Large values favour very sparse interaction networks, and the corresponding models will poorly fit the data. Conversely, for low $\gamma$, most of the interactions $T_{ij}$ are non-zero, and the corresponding model may be too flexible and overfit the data. In between the most adequate value of $\gamma$ can be obtained through the cross-validation procedure explained in section 4.1.5. A part of the data, the so-called training set $Y_{train}$, can be used to estimate the precision matrix $T^{MAP}(\gamma)$ for different regularisation strengths $\gamma$ through the maximisation of $\mathcal{L}(T, C_{train} = \frac{1}{M} Y_{train}^T \cdot Y_{train}, \gamma)$. The log-likelihood of the remaining data, gathered in the test (or validation) set $Y_{test}$ and defined by $\mathcal{L}(T^{MAP}(\gamma), C_{test} = \frac{1}{M} Y_{test}^T \cdot Y_{test})$, see Eq. (5.3), is then plotted as a function of $\gamma$. Its maximum locates the best hyper-parameter $\gamma$, see figure 4.2.

### 5.1.3   $L_1$-norm regularisation: effect of sampling noise

We now discuss the quality of the g-lasso inference as a function of the number of data $M$. For this, we consider as in Chapter 3 a null model with $L$ independent Gaussian variables $y$ associated to the precision matrix $T = \mathcal{I}_L$. We then collect $M$ independent realisations $y^m$, and consider the empirical covariance matrix $C$. The mean and the variance of the entry $C_{ij}$ (for $i < j$) defined in Eq. (3.17) are given by

$$\langle C_{ij} \rangle = 0 , \qquad \langle C_{ij}^2 \rangle = \frac{1}{M} , \qquad i < j . \tag{5.11}$$

We can also compute the covariance of two coefficients, with the result that, for $i < j$ and $k < l$, $\langle C_{ij} C_{kl} \rangle = 0$ unless $i = k$ and $j = l$. Note that $C_{ij}$ is symmetric by definition, so $C_{ij} = C_{ji}$ are obviously completely correlated. Hence, distinct matrix elements with $i < j$ have zero covariance. While in principle this does not mean that they are uncorrelated, we are going to neglect all higher-order correlations, which is correct when $M \gg L$. We thus approximate the matrix elements $C_{ij}$ with $i < j$ with independent random Gaussian variables of zero mean and variance $1/M$.

Let us estimate the value of the largest off-diagonal matrix element in this null model,

$$C_{noise} = \max_{i<j} |C_{ij}| . \tag{5.12}$$

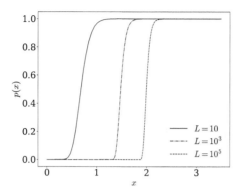

**Fig. 5.2** Probability $p(x)$ that the largest empirical covariance is at least equal to $x$, obtained from Eq. (5.13). From left to right: $L = 10$, $L = 1000$, $L = 100000$. For all three curves $M = 10$. Notice the steeper increase of $p$ for larger $L$.

Its value sets a lower threshold for extracting true correlations in real data sets. As finite sampling can produce fictitious correlations of magnitude $C_{noise}$, we should not expect that lower correlations have any particular meaning and are informative about true dependencies between variables.

The calculation of $C_{noise}$ is easily done via extreme value statistics. The probability $p(x)$ that $C_{noise}$ is smaller than some value, say, $x$ is equal to the probability that all $\frac{1}{2}L(L-1)$ correlations are smaller than $x$. We have

$$p(x) = \left( \int_{-\infty}^{x} dC \, \frac{e^{-MC^2/2}}{\sqrt{2\pi/M}} \right)^{\frac{L(L-1)}{2}} = \left( 1 - \int_{x\sqrt{M}}^{\infty} du \, \frac{e^{-u^2/2}}{\sqrt{2\pi}} \right)^{\frac{L(L-1)}{2}} . \tag{5.13}$$

The plot of $p(x)$ is shown in figure 5.2 for different values of $L$ and $M$. We observe the presence of a steep increase from $p \simeq 0$ to $p \simeq 1$ as $x$ crosses a critical value. This sharp transition implies that $C_{noise}$ takes, with high probability, a well-defined value that depends on $L$ and $M$. To determine this value we first write that, if $M, L, x\sqrt{M}$ are large, the previous expression for $p(x)$ can be approximated by

$$p(x) \sim \exp\left( -\frac{L^2}{2} \int_{x\sqrt{M}}^{\infty} du \, \frac{e^{-u^2/2}}{\sqrt{2\pi}} \right) \sim \exp\left( -\frac{L^2 \, e^{-Mx^2/2}}{2 \, x\sqrt{2\pi M}} \right) . \tag{5.14}$$

We then look for the root of $p(C_{noise}) = \frac{1}{2}$, and obtain[3]

$$C_{noise} \simeq 2 \sqrt{\frac{\log L}{M}} , \tag{5.15}$$

[3] The derivative of the cumulative probability $p(x)$ in $x = C_{noise}$ is of the order of $\sqrt{M \log L}$. This large slope ensure that choosing $\frac{1}{2}$ does not affect the leading order behaviour of $C_{noise}$ in Eq. (5.15). Note also that $C_{noise}\sqrt{M} \gg 1$, which justifies a posteriori the derivation of the asymptotic expression of $p(x)$ in Eq. (5.14).

up to additive corrections of the order of $\log \log L / \sqrt{M \log L}$. So, we conclude that the largest matrix element of a correlation matrix $C$ built by pure noise, in the limit of large $L$ and $M$, is $2\sqrt{\log L / M}$.

Based on Eq. (5.10), we conclude that while the true off-diagonal matrix elements $\widehat{T}_{ij}$ vanish, g-lasso would return non-zero matrix elements as soon as $\gamma < C_{noise}$. Therefore, Eq. (5.15) sets the minimal value of the regularisation strength $\gamma$ capable of removing the sampling noise in the correlation matrix. The result in Eq. (5.15) is encouraging, as it implies that, as far as network reconstruction is concerned, the number $M$ of data necessary to obtain some information about a problem with $L$ variables scales as $M \sim \log L$. This slow growth has to be compared with that of Chapter 3, where we discussed the inference of the eigenvalues of $T$. There, we found that we needed as many as $M \sim L$ data in order to correctly infer the top eigenvector of the covariance matrix. The substantial difference between these two scalings stems for the fact that eigenvectors are collective properties of the covariance matrix, which depend on all of its entries.

## 5.2  Boltzmann machines

In many applications, the distribution of variables are not Gaussian. In particular, in many biological applications, variables are categorical. For example, in neuron recordings a natural representation is binary (active or silent neuron), in DNA or RNA sequences variables take one of the 4 nucleotide values, and in protein sequences variables have 20 states, one for each amino acid. To avoid confusion with other types of continuous or discrete variables (denoted so far with $y$), we will use in the following the notation $\sigma$ to denote a $q$-state categorical variable; for simplicity we will often write $\sigma = 0, 1, \ldots, q - 1$, but no *a priori* order relation exists between the categories. As usual we will consider a configuration of $L$ such variables, $\boldsymbol{\sigma} = (\sigma_1, \cdots, \sigma_L)$. A standard approach to deal with the inference of distributions over such configurations is to use Boltzmann machines (BM).

### 5.2.1  No interaction: independent-site or profile model

We will start with the simplest case of independent variables, in which the distribution $p(\boldsymbol{\sigma})$ is the product of the single-variable distributions,

$$p_i(\sigma_i) = \frac{e^{h_i(\sigma_i)}}{\displaystyle\sum_{\sigma=0}^{q-1} e^{h_i(\sigma)}} \ , \qquad i = 1, \ldots, L \ . \tag{5.16}$$

The function $h_i(\sigma_i)$ plays the role of the log-likelihood (up to an additive constant) in statistics, of the local field in statistical mechanics, and of the position weight matrix in bioinformatics.

As $h_i$ can take $q$ different values (one for each $\sigma_i$) it offers an overparametrised representation of the distribution $p_i$. Due to the normalisation condition, only $q - 1$ independent parameters are needed. It is easy to check that $p_i$ is unchanged under the addition of an arbitrary (variable-independent) quantity $\delta h_i$ to the fields,

$h_i(\sigma_i) \to h_i(\sigma_i) + \delta h_i$. This property, which implies that the probability is invariant under a local transformation, is reminiscent of gauge invariance in physics; it implies that a model will be identifiable from data up to gauge invariance only.

Let us now see how these parameters can be inferred through maximum likelihood estimation. For notational simplicity we restrict to the case $q = 2$, corresponding to binary variables $\sigma_i \in \{0, 1\}$. Then

$$p_i(\sigma_i) = \frac{e^{h_i(\sigma_i)}}{e^{h_i(0)} + e^{h_i(1)}} = \frac{e^{h_i(\sigma_i) - h_i(0)}}{1 + e^{h_i(1) - h_i(0)}} = \frac{e^{h_i \sigma_i}}{1 + e^{h_i}}, \tag{5.17}$$

where $h_i \equiv h_i(1) - h_i(0)$ is the gauge-invariant field acting on the variable $\sigma_i$, now taken as a true 0 or 1 number (rather than a category). We call $\boldsymbol{h} = (h_1, \cdots, h_L)$ the set of these field parameters.

We now randomly draw $M$ configurations $\boldsymbol{\sigma}^m$, $m = 1, \ldots, M$. The log-likelihood per data point reads

$$\mathcal{L}(\boldsymbol{h}) = \frac{1}{M} \sum_{m=1}^{M} \log p(\boldsymbol{\sigma}^m) = \frac{1}{M} \sum_{i=1}^{L} \sum_{m=1}^{M} \log p_i(\sigma_i^m) = \sum_{i=1}^{L} \left( h_i \mu_i - \log(1 + e^{h_i}) \right), \tag{5.18}$$

where

$$\mu_i = \frac{1}{M} \sum_{m=1}^{M} \sigma_i^m \tag{5.19}$$

is the empirical average of variable $\sigma_i$. The MLE estimator is obtained by maximising the log-likelihood, which gives

$$\frac{\partial}{\partial h_i} \mathcal{L}(\mathbf{h}) = 0 = \mu_i - \frac{e^{h_i}}{1 + e^{h_i}}. \tag{5.20}$$

As $e^{h_i}/(1 + e^{h_i}) = 0 \times p_i(0) + 1 \times p_i(1)$ is the average value of $\sigma_i$ according to the model distribution, this equation is called "moment-matching" condition: it demands that the empirical and model averages coincide. The solution to Eq. (5.20) is obviously unique. Notice that it is easy to check that the second derivative of $\mathcal{L}$ is strictly negative, and, hence, that $\mathcal{L}$ is concave.

To conclude, let us define the empirical distribution over $\sigma_i$ through

$$p_i^{emp}(\sigma_i) = \frac{1}{M} \sum_{m=1}^{M} \delta_{\sigma_i, \sigma_i^m}, \qquad \mu_i = \sum_{\sigma_i} \sigma_i \, p_i^{emp}(\sigma_i). \tag{5.21}$$

We may rewrite the log-likelihood as

$$\mathcal{L}(\boldsymbol{h}) = \sum_{i=1}^{L} \sum_{\sigma_i = 0,1} p_i^{emp}(\sigma_i) \log p_i(\sigma_i), \tag{5.22}$$

which is minus the cross entropy between the empirical and model distributions. At the maximum of the log-likelihood (corresponding to the minimum of the cross entropy), both distributions become equal, and the log-likelihood is then equal to minus their entropy.

We can compare this with the maximum entropy principle that was discussed in Chapter 2. The entropy of an arbitrary distribution, constrained to be normalised and to match the empirical averages of the variables $\sigma_i$, is

$$S[p] = -\sum_{\sigma} p(\boldsymbol{\sigma}) \log p(\boldsymbol{\sigma}) + \lambda \left( \sum_{\sigma} p(\boldsymbol{\sigma}) - 1 \right) + \sum_{i=1}^{L} \lambda_i \left( \sum_{\sigma} \sigma_i \, p(\boldsymbol{\sigma}) - \mu_i \right) , \quad (5.23)$$

where the $\lambda$'s are Lagrange multipliers enforcing the constraints. Maximisation over $p(\boldsymbol{\sigma})$ produces exactly the same result as that obtained above, with $\lambda_i = h_i$.

## 5.2.2   Pairwise interactions: Boltzmann machine

Consider now the case of a model over the configurations $\boldsymbol{\sigma}$, with interdependencies between the variables. In the case of pairwise interactions only[4] we write the distribution as

$$p(\boldsymbol{\sigma}) = \frac{1}{Z(\boldsymbol{h}, \boldsymbol{J})} \prod_{i=1}^{L} e^{h_i(\sigma_i)} \prod_{i<j} e^{J_{ij}(\sigma_i, \sigma_j)} = \frac{1}{Z(\boldsymbol{h}, \boldsymbol{J})} e^{-E(\boldsymbol{\sigma}; \boldsymbol{h}, \boldsymbol{J})} , \quad (5.24)$$

where the "energy" function,

$$E(\boldsymbol{\sigma}; \boldsymbol{h}, \boldsymbol{J}) = -\sum_{i} h_i(\sigma_i) - \sum_{i<j} J_{ij}(\sigma_i, \sigma_j) , \quad (5.25)$$

is reminiscent of that of Potts models in statistical mechanics. The partition function,

$$Z(\boldsymbol{h}, \boldsymbol{J}) = \sum_{\sigma} e^{-E(\boldsymbol{\sigma}; \boldsymbol{h}, \boldsymbol{J})} , \quad (5.26)$$

is generally intractable, as its computation requires to sum over $q^L$ states of the variables.

Notice that the gauge invariance we discussed in the case of the independent-site model is now extended to the pair interactions. It is easy to check that the distribution is left unchanged under all the transformations of the type

$$J_{ij}(\sigma_i, \sigma_j) \to J_{ij}(\sigma_i, \sigma_j) + \delta K_{ij}(\sigma_i) + \delta \tilde{K}_{ij}(\sigma_j) ,$$
$$h_i(\sigma_i) \to h_i(\sigma_i) - \sum_{j(>i)} \delta K_{ij}(\sigma_i) - \sum_{j(<i)} \delta \tilde{K}_{ji}(\sigma_i) + \delta h_i . \quad (5.27)$$

As a result, a lot of concerted changes to the couplings and the fields are possible.

---

[4]Additional terms in the energy involving triplets, quadruplets, etc., of variables could of course be included, but for the moment we will consider only pair interactions that already contain a lot of additional information with respect to the independent-site model. We will come back on multi-spin interactions in section 6.2.

Let us again turn to the case $q = 2$ for the sake of simplicity, with $\sigma_i \in \{0, 1\}$ variables. With a proper choice of gauge, the energy function is now

$$E(\sigma; \boldsymbol{h}, \boldsymbol{J}) = -\sum_i h_i \, \sigma_i - \sum_{i<j} J_{ij} \, \sigma_i \, \sigma_j \; . \tag{5.28}$$

The log-likelihood (or, equivalently, minus the cross entropy) per data point reads

$$\mathcal{L}(\boldsymbol{h}, \boldsymbol{J}) = \sum_i h_i \, \mu_i + \sum_{i<j} J_{ij} \, \mu_{ij} - \log Z(\boldsymbol{h}, \boldsymbol{J}) \; , \tag{5.29}$$

where $\mu_i$ is the empirical mean of $\sigma_i$, see Eq. (5.19), and $\mu_{ij}$ is the empirical mean of $\sigma_i \sigma_j$. Differentiating $\mathcal{L}(\boldsymbol{h}, \boldsymbol{J})$ with respect to the fields and the couplings, we obtain moment-matching conditions between the one- and two-point statistics of the data and of their model counterparts,

$$\frac{\partial}{\partial h_i} \mathcal{L}(\mathbf{h}) = 0 = \mu_i - \frac{\partial}{\partial h_i} \log Z(\boldsymbol{h}, \boldsymbol{J}) = \mu_i - \langle \sigma_i \rangle \; ,$$

$$\frac{\partial}{\partial J_{ij}} \mathcal{L}(\mathbf{h}) = 0 = \mu_{ij} - \frac{\partial}{\partial J_{ij}} \log Z(\boldsymbol{h}, \boldsymbol{J}) = \mu_{ij} - \langle \sigma_i \sigma_j \rangle \; , \tag{5.30}$$

where

$$\langle \sigma_i \rangle = \sum_\sigma \sigma_i \, p(\boldsymbol{\sigma}) \; , \qquad \langle \sigma_i \sigma_j \rangle = \sum_\sigma \sigma_i \, \sigma_j \, p(\boldsymbol{\sigma}) \; , \tag{5.31}$$

are the first and second moments computed with the model distribution $p(\boldsymbol{\sigma})$ in Eq. (5.24). As in the case of the independent-site model, one can easily show that the log-likelihood is a concave function, and has therefore a unique maximum.

The $\frac{1}{2}L\,(L{+}1)$ moment-matching conditions in Eqs. (5.30) must be solved to obtain the $L$ fields $h_i$ and $\frac{1}{2}L\,(L-1)$ couplings $J_{ij}$. Note that, with limited sampling, it is possible that some empirical moments take boundary values, *e.g.* $\mu_i = 1$ or $\mu_{ij} = 0$, which require, respectively, infinite fields $h_i = +\infty$ or couplings $J_{ij} = -\infty$. These divergences can be avoided by including a regularisation in the inference, *i.e.* a prior in the Bayesian inference setting, or a pseudo-count in the calculation of the empirical averages.

### 5.2.3  Boltzmann machine learning

Solving directly the moment-matching Eqs. (5.30) is a highly non-trivial numerical task, except in the case of the independent-site model, or for very peculiar forms of the coupling matrix, due to the difficulty in computing averages over the model distribution. Monte Carlo simulation methods for fixed $\boldsymbol{h}, \boldsymbol{J}$ can, however, be applied. A flexible numerical method, known under the name of Boltzmann machine learning method [25], consists in alternating accurate Monte Carlo-based estimation of the one- and two-points moments, as well as changes of the parameters $\boldsymbol{h}$ and $\boldsymbol{J}$. It can be summarised as follows:

---

**Boltzmann machine learning algorithm**

- Start from random (or educated, *e.g.* $\boldsymbol{J} = 0$ and the solutions of Eqs. (5.20) for $\boldsymbol{h}$) guesses for $\boldsymbol{h}$ and $\boldsymbol{J}$.
- Run Monte Carlo simulations with energy $E(\boldsymbol{\sigma}; \boldsymbol{h}, \boldsymbol{J})$ to estimate the moments $\langle \sigma_i \rangle$ and $\langle \sigma_i \sigma_j \rangle$.
- Update the coupling and fields through

$$
\begin{aligned}
h_i^{t+1} &= h_i^t + \eta_h \frac{\partial \mathcal{L}}{\partial h_i}(\boldsymbol{h}^t, \boldsymbol{J}^t) = h_i^t + \eta_h \left( \mu_i - \langle \sigma_i \rangle \right), \\
J_{ij}^{t+1} &= J_{ij}^t + \eta_J \frac{\partial \mathcal{L}}{\partial J_{ij}}(\boldsymbol{h}^t, \boldsymbol{J}^t) = J_{ij}^t + \eta_J \left( \mu_{ij} - \langle \sigma_i \sigma_j \rangle \right),
\end{aligned}
\tag{5.32}
$$

where $\eta_h, \eta_J$ are adequately chosen learning rates.

---

The intuition underlying this update rule is that, if for example the average of $\sigma_i$ in the model is larger than in the data, the value of $h_i$ should be decreased to match these two moments.

As these equations correspond to a gradient ascent dynamics for the log-likelihood, see Eq. (5.30) (equivalently, to gradient descent of the cross entropy), convergence towards the maximum is guaranteed, provided $\eta_{h,J}$ are diminished and eventually sent to zero to gently stop in the maximum. However, Boltzmann machine learning suffers from some drawbacks.

First, as the learning dynamics is formulated in a very high-dimensional space $\{\boldsymbol{h}, \boldsymbol{J}\}$, the log-likelihood could be extremely steep along some directions and very flat along others. Thus, a single learning rate $\eta_J$ for the couplings (the problem is usually milder with the fields) could be either too large, leading to overshooting along the steeper directions, or too small, yielding no parameter updates along the flat directions. This issue could in principle be solved with second-order methods, a generalisation of the Newton-Raphson method. However estimating the curvature matrix is quite complex and time-demanding in high dimensions [26].

Second, Boltzmann machine learning may require massive Monte Carlo sampling, because the moments should be computed at each update step. In each Monte Carlo simulation, one should be sure that equilibration has been realised and that the moment are correctly estimated. An alternative approach is contrastive divergence (CD) [27], which consists in replacing the exact model averages in Eq. (5.32) by approximations, $\langle \bullet \rangle \rightarrow \langle \bullet \rangle_{\mathrm{CD}}$ computed as follows. One initialises $M$ running configurations $\boldsymbol{\sigma}^m$ to the values of the $M$ available data points. Then, at each iteration, one performs $k$ Monte Carlo step to update each one of the $M$ configurations, and computes $\langle \bullet \rangle_{\mathrm{CD}}$ as an empirical average over the running configurations. Fields and couplings are then updated according to Eq. (5.32).

CD is an empirical method, but the rationale behind it is that when the gradient of the log-likelihood becomes small, the model distribution is rather close to the empirical one. So, even if at each step the Monte Carlo chain is not guaranteed to be equilibrated (to the current model distribution), all distributions will converge in the long run, and the simulation will equilibrate. In many practical cases, CD may work quite well even with $k = 1$, see [28] for a detailed discussion of the limits of this approach. Note that one can also decide to use mini-batches within CD, in order to speed up the evaluation of the empirical and running averages.

### 5.2.4 Mean-field inference

It may be useful for some applications to have approximate analytical methods to estimate the couplings and fields. Mean-field theory offers such a possibility. It starts from the so-called Callen identities, which read, for $\sigma_i \in \{0,1\}$,

$$\langle \sigma_i \rangle = \left\langle F\left(h_i + \sum_j J_{ij}\,\sigma_j\right)\right\rangle \quad \text{with} \quad F(x) = \frac{e^x}{1 + e^x}\,, \tag{5.33}$$

where the averages are taken over the model distribution. This identity can be rigorously established[5], and intuitively understood as follows. Draw a configuration $\sigma$ from the model distribution $p(\sigma)$. Then freeze all variables to their values $\sigma_j$ but one, say, $i$, and draw it repeatedly from its conditional probability given the other, frozen variables. The average value of $\sigma_i$ in the background of the frozen variable is $F(h_i + \sum_j J_{ij}\,\sigma_j)$. Now we can release all variables, draw another configuration $\sigma$ and repeat this process. We end up with Eq. (5.33).

Note that the couplings were originally defined in Eq. (5.24) only for ordered pairs of indices $i < j$; here we introduce $J_{ji} = J_{ij}$ and assume $J_{ii} = 0$ to lighten the notation. The mean-field approximation consists in moving the average inside the argument of $F$,

$$\langle \sigma_i \rangle_{mf} = F\left(h_i + \sum_j J_{ij}\,\langle \sigma_j \rangle_{mf}\right). \tag{5.34}$$

This approximation can be expected to be better if the connectivity of a variable (the number of interactions it depends on) is large, as the argument of $F$ is the sum of a large number of terms and may be closer to its mean. It defines a set of $L$ coupled equations over the $L$ mean-field first moments $\langle \sigma_i \rangle_{mf}$.

Let us now imagine we perturb a little bit the field $h_i$ acting on site $i$. This will result in a direct change of the mean-field average $\langle \sigma_i \rangle_{mf}$ through Eq. (5.34), which in turn will produce indirect changes of the other first moments $\langle \sigma_j \rangle_{mf}$, with $j \neq i$. We characterise this changes through the linear response matrix

$$R_{ij} = \frac{\partial \langle \sigma_i \rangle}{\partial h_j}\,. \tag{5.35}$$

We can compute $R$ in two ways:

[5] For any configuration we write $\sigma = (\hat{\sigma}_i, \sigma_i)$, with $\hat{\sigma}_i = \{\sigma_j, j \neq i\}$. Then,

$$\langle \sigma_i \rangle = \sum_\sigma \sigma_i\, p(\sigma) = \sum_{\hat{\sigma}} p(\hat{\sigma}) \sum_{\sigma_i=0,1} \sigma_i\, p(\sigma_i|\hat{\sigma}) = \sum_{\hat{\sigma}} p(\hat{\sigma})\, F\left(H_i(\hat{\sigma})\right),$$

where $-\sigma_i H_i(\hat{\sigma}) = -\sigma_i(h_i + \sum_j J_{ij}\,\sigma_j)$ is the $\sigma_i$-dependent contribution to the energy $E(\hat{\sigma}_i, \sigma_i)$.

- First we differentiate the mean-field Eqs. (5.34) with respect to $h_j$, to obtain

$$R_{ij} = D_{ii} \left\{ \delta_{ij} + \sum_k J_{ik} R_{kj} \right\} , \quad \text{with } D_{ii} \equiv F' \left( h_i + \sum_j J_{ij} \langle \sigma_j \rangle_{mf} \right) . \quad (5.36)$$

We now consider the diagonal matrix $D$, whose diagonal elements are the $D_{ii}$'s defined above. The previous equation can be written in matrix form as $R = D \cdot (\mathcal{I}_L + J \cdot R)$, which yields

$$J = D^{-1} - R^{-1} . \quad (5.37)$$

- Second, the response matrix obeys the general fluctuation-dissipation relation, see Eqs. (5.25) and (5.26),

$$R_{ij} = \frac{\partial}{\partial h_j} \frac{\partial}{\partial h_i} \log Z(\boldsymbol{h}, \boldsymbol{J}) = \langle \sigma_i \sigma_j \rangle - \langle \sigma_i \rangle \langle \sigma_j \rangle = C_{ij} , \quad (5.38)$$

where $C_{ij}$ is the connected correlation matrix.

We may now use the fluctuation-dissipation relation $C = R$ in Eq. (5.37) for the coupling matrix. As $D$ is a diagonal matrix, we end up with

$$J_{ij} = -C_{ij}^{-1} , \quad (5.39)$$

where the matrix $C_{ij}$ can now be estimated directly as an empirical average over the data. The outcome of the mean-field computation is therefore equivalent to what we obtained for multivariate Gaussian variables through maximum likelihood estimation.

## 5.3 Pseudo-likelihood methods

Mean-field inference is fast but suffers from a major drawback: it is not statistically consistent. In other words, if we generate data from a graphical model, the ground truth interactions are not exactly recovered even in the limit of infinite sampling ($M \to \infty$ at fixed $L$). Hereafter, we introduce another approach, called pseudo-likelihood maximisation (PLM), which is very efficient from a computational point of view, and, yet, is statistically consistent. Informally speaking, PLM is based on the idea of inferring the interaction networks row by row, one at a time. This strategy is much more effective than inferring the whole network at once, as we have done so far.

Given a configuration $\boldsymbol{y}$ of the variables and an index $i = 1, \dots, L$, we define $\boldsymbol{y}_{\backslash i}$ the set of all variables except $y_i$, as in section 5.1. The pseudo-likelihood of a configuration $\boldsymbol{y}$ is defined as the product of the conditional probabilities of all its variables,

$$\mathcal{PL}(\boldsymbol{y}|\boldsymbol{\theta}) = \prod_{i=1}^{L} p(y_i|\boldsymbol{y}_{\backslash i}, \boldsymbol{\theta}) , \quad (5.40)$$

and the pseudo-likelihood $\mathcal{PL}(Y|\boldsymbol{\theta})$ of independently drawn data points $Y = \{\boldsymbol{y}^m\}, m = 1, \dots, M$, is simply the product of the pseudo-likelihoods of every configuration $\boldsymbol{y}^m$. PLM amounts to infer $\boldsymbol{\theta}$ from the maximisation of $\mathcal{PL}(Y|\boldsymbol{\theta})$.

The pseudo-likelihood is not equal to the likelihood, except for the case of independent variables. However, as we will see below, PLM gives back the right interaction network when an infinite number of data are available.

### 5.3.1 Case of multivariate Gaussian variables: perfect sampling

Consider the case of data $Y$ generated from a centred multivariate Gaussian distribution with precision matrix $\widehat{T}$. The pseudo-loglikelihood for $Y$ and a precision matrix $T$ is easily expressed from the conditional probabilities given in Eq. (5.1). It reads, up to irrelevant additive terms,

$$\log \mathcal{PL}(Y|T) = \sum_{m,i} \log p\left(y_i^m \middle| y_{\backslash i}^m, T\right) = \frac{M}{2} \sum_i \log T_{ii} - \frac{M}{2} \sum_i T_{ii} \, G_i\left(Y, T_{i\bullet}\right) , \quad (5.41)$$

where $T_{i\bullet}$ denotes the $i^{th}$ row of the precision matrix, *i.e.* the set of interactions $T_{ij}$ with $i$ fixed and $j \neq i$, and

$$G_i\left(Y, T_{i\bullet}\right) = \frac{1}{M} \sum_m \left( y_i^m + \sum_{j(\neq i)} \frac{T_{ij}}{T_{ii}} y_j^m \right)^2 . \quad (5.42)$$

We observe that the different rows are decoupled in Eq. (5.41), *i.e.* we can maximise the pseudo-likelihood over each row, independently of the others. Obviously, the inferred precision matrix will generally not be symmetric: we will come back to this point in the next section.

Let us now assume that we have a perfect sampling ($M \to \infty$). Expanding the quadratic form defined by $G$ and denoting by $\widehat{C} = \widehat{T}^{-1}$ the covariance matrix of the distribution from which the data were drawn, we obtain

$$G_i(t) = \widehat{C}_{ii} + 2 \sum_{j(\neq i)} t_j \, \widehat{C}_{ij} + \sum_{j,k(\neq i)} t_j \, t_k \, \widehat{C}_{jk} , \quad (5.43)$$

where $t_j = T_{ij}/T_{ii}$. To maximise the pseudo-loglikelihood over the row $T_{i\bullet}$, we minimise $G_i$ over the variables $t_j$:

$$\left. \frac{\partial G_i}{\partial t_j} \right|_{t^{PLM}} = 0 = 2\, \widehat{C}_{ij} + 2 \sum_{k(\neq i)} \widehat{C}_{jk} \, t_k^{PLM} \quad \Rightarrow \quad t_j^{PLM} = - \sum_{k(\neq i)}^{L} \left[\widehat{C}_R^{-1}\right]_{kj} \widehat{C}_{ji} . \quad (5.44)$$

In the equation above $\widehat{C}_R$ denotes the $(L-1)$-dimensional restriction of $\widehat{C}$ obtained upon removal of row and column $i$:

$$\widehat{C} = \begin{pmatrix} \widehat{C}_{ii} & \widehat{C}_{i\bullet} \\ \widehat{C}_{\bullet i} & \widehat{C}_R \end{pmatrix} , \quad (5.45)$$

where we have permuted the indices to simplify the block representation of the covariance matrix. The inverse of $\widehat{C}_R$, which appears in Eq. (5.44) can be expressed using general block-inversion formulae for matrices:

$$\widehat{C}^{-1} = \frac{1}{A} \begin{pmatrix} 1 & -\widehat{C}_{i\bullet} \cdot \widehat{C}_R^{-1} \\ -\widehat{C}_R^{-1} \cdot \widehat{C}_{\bullet i} & A\,\widehat{C}_R^{-1} + \widehat{C}_R^{-1} \cdot \widehat{C}_{\bullet i} \cdot \widehat{C}_{i\bullet} \cdot \widehat{C}_R^{-1} \end{pmatrix} , \quad (5.46)$$

where $A = \widehat{C}_{ii} - \widehat{C}_{i\bullet} \cdot \widehat{C}_R^{-1} \cdot \widehat{C}_{\bullet i}$.

Looking at the first column of the block representation of $\widehat{C}^{-1}$ we see that the PLM estimator found in Eq. (5.44) is equivalent to

$$t_j^{PLM} = A \, [\widehat{C}^{-1}]_{ij} = \frac{[\widehat{C}^{-1}]_{ij}}{[\widehat{C}^{-1}]_{ii}} = \frac{\widehat{T}_{ij}}{\widehat{T}_{ii}} \quad \Rightarrow \quad T_{ij}^{PLM} = \widehat{T}_{ij} \times \frac{T_{ii}^{PLM}}{\widehat{T}_{ii}} \, . \tag{5.47}$$

We are left with the inference of the diagonal element of the precision matrix with PLM. To do so, we maximise Eq. (5.41) over $T_{ii}$, with the result

$$T_{ii}^{PLM} = \frac{1}{G_i(t^{PLM})} = \frac{\left([\widehat{C}^{-1}]_{ii}\right)^2}{\left(\frac{[\widehat{C}^{-1}]_{ii}}{[\widehat{C}^{-1}]_{\bullet i}}\right)^T \cdot \widehat{C} \cdot \left(\frac{[\widehat{C}^{-1}]_{ii}}{[\widehat{C}^{-1}]_{\bullet i}}\right)} = [\widehat{C}^{-1}]_{ii} = \widehat{T}_{ii} \, . \tag{5.48}$$

As a conclusion, we have shown that, for perfect sampling, PLM gives back the ground-truth precision matrix. This property is not specific to multivariate Gaussian distributions, but can be proven for generic distributions [29].

### 5.3.2 Case of multivariate Gaussian variables: limited sampling

In practical applications, sampling is of course never perfect. For finite $M$, it is natural to introduce a modified version of the cost function $G$ introduced in Eq. (5.42), regularised with a $L_1$-norm log prior,

$$G^{L_1}(Y, t, \gamma) = \frac{1}{M} \sum_m \left( y_i^m + \sum_{j(\neq i)} t_j \, y_j^m \right)^2 + \gamma \sum_{j(\neq i)} |t_j| \, . \tag{5.49}$$

This cost function is a convex function of its arguments $t_j$ and can be minimised efficiently. It is possible to show [30] that, if the number $M$ of data is large compared to $\log L$ and the regularisation parameter is chosen to be $\gamma \gtrsim \sqrt{\log L/M}$, then the entries $T_{ij}$ of the precision matrix larger than $M^{-1/2} \sim (\log L)^{-1/2}$ can be identified as different from zero with high probability. As a consequence, the support of the graph of dependencies can be efficiently recovered.

This procedure is computationally very efficient, but it has the disadvantage that the estimate of $T_{ij}$ can be different from that of $T_{ji}$. To enforce the symmetry of the precision matrix, a simple but computationally heavy procedure is to consider the total pseudo-loglikelihood, $\log \mathcal{PL}(Y|T)$, where all elements $T_{ij}$ with $i > j$ are replaced by $T_{ji}$. The resulting expression, added to the $L_1$-norm log prior, is then maximised over the upper triangular matrix $T_{ij}$ with $i \leq j$. The rows of $T$ are not uncoupled any longer, but the pseudo-loglikelihood is still convex and can be maximised with appropriate numerical procedures.

### 5.3.3 Case of categorical variables

The PLM procedure presented above can be adapted to non-Gaussian distributions, *e.g.* defined on categorical variables. For simplicity we consider the case of $q = 2$ categories, denoted by $\sigma = 0, 1$, in the following. Let $\boldsymbol{\theta} = \{h_i, J_{ij}\}$, $i, j = 1, \ldots, L$ be the set of all fields and couplings, and $\Sigma$ a $M \times L$ matrix containing $M$ data points $\{\boldsymbol{\sigma}^m\}, m = 1, \ldots, M$.

Let $\boldsymbol{\sigma}_{\backslash i}$ be the configuration of variables except $\sigma_i$. The conditional probability of the $i^{th}$ variable reads

$$p(\sigma_i|\boldsymbol{\sigma}_{\backslash i}, h_i, J_{i\bullet}) = \frac{e^{H_i \sigma_i}}{1 + e^{H_i}} \quad \text{with} \quad H_i(\boldsymbol{\sigma}_{\backslash i}, h_i, J_{i\bullet}) = h_i + \sum_{j(\neq i)} J_{ij} \sigma_j \,. \tag{5.50}$$

Notice that $H_i$ coincides with the argument of the sigmoidal function $F$ in Eq. (5.33). We then write the contribution to the pseudo-loglikelihood associated to the $i^{th}$ row of the coupling matrix,

$$\log \mathcal{PL}_i(\Sigma, h_i, J_{i\bullet}) = \sum_{m=1}^{M} \log p(\sigma_i^m|\boldsymbol{\sigma}_{\backslash i}^m, h_i, J_{i\bullet}) \tag{5.51}$$

$$= \sum_{m} \left[\sigma_i^m H_i(\boldsymbol{\sigma}_{\backslash i}^m, h_i, J_{i\bullet}) - \log\left(1 + e^{H_i(\boldsymbol{\sigma}_{\backslash i}^m, h_i, J_{i\bullet})}\right)\right].$$

This quantity is, for each $i$, a concave function of its arguments $h_i$ and $J_{i\bullet}$ and can be maximised to infer their values. At the maximum the following moment-matching conditions are met,

$$\frac{1}{M} \sum_{m} \sigma_i^m = \frac{1}{M} \sum_{m} F\left(H_i(\boldsymbol{\sigma}_{\backslash i}^m, h_i, J_{i\bullet})\right) \,,$$

$$\frac{1}{M} \sum_{m} \sigma_i^m \sigma_j^m = \frac{1}{M} \sum_{m} F\left(H_i(\boldsymbol{\sigma}_{\backslash i}^m, h_i, J_{i\bullet})\right) \sigma_j^m \,, \tag{5.52}$$

where $F(x) = e^x/(1+e^x)$. These conditions generalise the Callen identity of Eq. (5.33), and impose the equality between the one- and two-point moments computed from the data and from a model where a single variable is subjected to its local field and the action of the other variables sampled from the data.

### 5.3.4 Concluding remarks

We end up this section on pseudo-likelihood with a few comments:

- As in the case of Gaussian variables, PLM for finite $M$ is an arbitrary approximation and it is not formulated in a Bayesian inference framework. However, when $M \to \infty$, the empirical averages converge to the exact ones. As PLM is equivalent to finding fields and couplings such that the exact Callen identities hold for the empirical averages, it provides the correct answer when $M \to \infty$.
- For finite $M$, maximisation of $\mathcal{PL}_i$ provides an estimate of $J_{i\bullet}$ while maximisation of $\mathcal{PL}_j$ provides an estimate of $J_{j\bullet}$. As a result, the inferred values of $J_{ij}$ and $J_{ji}$ will generally differ. To enforce symmetry, as discussed in section 5.3.2, one can maximise the total pseudo-loglikelihood, $\sum_i \log \mathcal{PL}_i$, which is, however, computationally more expensive.

- From the algorithmic point of view, PLM has the advantage, with respect to MLE, to offer independent maximisation problems over the rows $J_{i\bullet}$. Furthermore, the evaluation of the gradient of $\mathcal{PL}_i$ only requires the evaluation of empirical means in Eq. (5.52), which avoids the need for Monte Carlo simulation as required in Boltzmann machine learning. Unfortunately, because the empirical means involving $H_i$ are non-linear functions of the couplings, they must be re-evaluated at each step of the maximisation procedure, which requires a time proportional to the number $M$ of data. Conversely, in regular Boltzmann machine learning, the empirical averages are calculated once at the beginning of the procedure. The use of mini-batches, *i.e.* the evaluation of the empirical averages on a smaller subset of the total available dataset, might be considered to obtain a speed-up.
- Last of all, as discussed in section 5.3.2, a regularisation can be added to the pseudo-likelihood in order to enforce sparsity and/or avoid overfitting.

## 5.4   Tutorial 5: inference of protein structure from sequence data

Protein folding is one of the most important problems in structural biology. It consists in predicting how a protein folds from the sole knowledge of its sequence. Thanks to great progress in sequencing technology over the last 20 years, databases such as PFAM [31] now contain large numbers of protein sequences, obtained from many organisms, as diverse as bacteria and eukaryots. Based on their similarities, sequences can be partitioned into protein families sharing a common ancestral origin. Proteins belonging to the same family share, as a rough approximation, the same structure and function. Their sequences may differ in length, due to the insertion and deletion processes that took place during evolution, and need to be aligned to a common format to allow for comparison[6]. They are then collected into what is called a multiple sequence alignment (MSA) of the protein family.

The goal of this tutorial is to extract from the MSA of the protein family PF00014 (trypsin inhibitor protein domain), downloaded from the PFAM database, the correlation matrix between amino acids on different sites, and to infer from such data an interaction network between the sites of the protein. As shown in figure 5.3, a physical hypothesis is that the sites that are nearby in the three-dimensional structure of the protein have strong interactions, and therefore the inferred interactions can be used as predictors of three-dimensional contacts between sites [34, 35]. This kind of analysis (known in the field as direct-coupling analysis, or DCA) are at the core of softwares like AlphaFold [36] and trRosetta [37], which recently revolutionised the field by providing extremely accurate structure predictions for individual natural protein sequences.

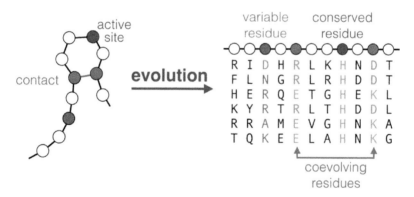

**Fig. 5.3** The structural and functional constraints acting on a protein limit the diversity of admissible sequences. In turn, the low-order statistics of amino acids in multi-sequence alignments of protein sequences coming from many evolutionary diverged organisms reflects those constraints. Figure reprinted from Ref. [34].

---

[6]Excellent introductory textbooks on sequence alignment exist [32, 33], so we do not discuss this problem in this book.

### 5.4.1 Preliminary calculations

Proteins are chains of amino acids. Each amino acid is defined by its position along the chain, called site, and its type, called residue. Data are the MSA for a protein family, in the format shown in figure 5.3. Typical MSA contain $M = 100 - 100000$ sequences of proteins with $L = 50 - 5000$ sites. Each row of the MSA is a single protein sequence, $a_1^m, \ldots, a_L^m$, with $m = 1, \ldots, M$, while each column represents a specific site in the protein chain. The alphabet of a protein sequence contains 20 possible letters indicating the residue type, plus the gap "-" symbol, which indicates a missing residue in a given sequence. This symbol is used to match sequences of different length in the MSA. Therefore each sequence is represented, on each of the $L$ residues, by a categorical variable with $q = 21$ symbols (or categories), leading to a $q$-state Potts model representation (section 5.2).

- **Frequencies** – From the MSA it is possible to extract the empirical frequencies $f_i(a)$ of occurrence of amino acid $a$ on a given site $i$, and the pairwise frequencies of occurrence of a pair of amino acids $a, b$ on a given pair of sites $i, j$, by simply counting their occurrences in the MSA. In formulae:

$$f_i(a) = \frac{1}{M} \sum_{m=1}^{M} \delta_{a_i^m, a} \ ,$$

$$f_{ij}(a, b) = \frac{1}{M} \sum_{m=1}^{M} \delta_{a_i^m, a} \delta_{a_j^m, b} \ , \tag{5.53}$$

where $\delta_{a,b}$ is the Kronecker delta, which equals one if $a = b$ and zero otherwise. Note that on the "diagonal blocks" where $i = j$ we trivially have $f_{ii}(a, b) = f_i(a)\delta_{a,b}$ because only one amino acid can be present on a site, and $f_{ij}(a, b) = f_{ji}(b, a)$, so one can restrict to $i < j$. Furthermore, the frequencies satisfy the relations

$$\sum_a f_i(a) = 1 \ ,$$

$$\sum_a f_{ij}(a, b) = f_j(b) \ , \qquad \sum_b f_{ij}(a, b) = f_i(a) \ . \tag{5.54}$$

- **Potts model** – The Boltzmann probability distribution of a Potts model

$$P(a_1, \ldots, a_L) = \frac{1}{Z} \exp\{-E(a_1, \ldots, a_L)\} \tag{5.55}$$

associates to each amino-acid sequence $a_1, \ldots, a_L$ its probability to be in the MSA of a given protein family. The energy of the model has the form:

$$E(a_1, \ldots, a_L) = - \sum_{1 \leq i < j \leq L} J_{ij}(a_i, a_j) - \sum_{i=1}^{L} h_i(a_i) \ . \tag{5.56}$$

The parameters of the model are the local fields $h_i(a)$ and the pairwise couplings $J_{ij}(a, b)$ for $i < j$. The fields and couplings have to be fitted to reproduce, from the equilibrium distribution Eq. (5.55) of the Potts model, configurations with the same statistical properties of the MSA, reproducing in particular the empirical frequencies and the pairwise correlations of Eq. (5.53).

- **Gauge invariance** – In Eq. (5.53) there are $q - 1$ independent amino-acid frequencies on each site, and $(q - 1)^2$ independent pairwise correlations due to the conditions expressed in Eq. (5.54). The Potts energy, which has $q$ field parameters for each site and $q^2$ coupling parameters for each pair of sites, is therefore overparametrised. As a result there exist local transformations of the parameters that keep the probability of a sequence in Eq. (5.55) unchanged. These transformations are listed in Eq. (5.27). In words, the transformation consists in adding, at fixed pair $(i < j)$ of sites, a constant to each row and column of the $q \times q$ couplings matrix. At fixed site $i$, one can add a constant to the field parameters, together with the terms compensating the coupling changes. The Ising case $q = 2$ is particularly useful for illustration. The inference procedure is usually carried out in a convenient gauge, see below.

- **Occurrence variables** – It is convenient to introduce occurrence variables $s_{ia} = \delta_{a_i,a}$ (also called a *one-hot encoding* of a categorical variable), which take the value 1 if the amino acid $a$ is present at position $i$, and zero otherwise. A sequence is therefore represented by a binary string of $Lq$ components instead of using a chain of $L$ symbols $a_i$ that can take each $q$ values. Note that the average of $s_{ia}$ over the empirical data is the occurrence frequency $f_i(a)$ of amino acid $a$ in position $i$.

- **Gaussian approximation** – The Gaussian approximation [35] consists in neglecting the discrete character of the occurrence variables $s_{ia}$. If these variables are replaced by $Lq$ Gaussian variables $y_{ia} = s_{ia} - f_i(a)$, centred around the empirical frequencies, the calculation of the partition function and the solution of the inverse problem is straightforward:

$$p(\mathbf{y}|J) = \frac{\sqrt{\det(-J)}}{(2\pi)^{(Lq)/2}} e^{\frac{1}{2}\sum_{ia,jb} y_{ia} J_{ij}(a,b) y_{jb}} . \tag{5.57}$$

As discussed in Chapter 3, the maximisation of the likelihood of the data (here, the MSA) is equivalent to the minimisation of the cross entropy,

$$S_c = -\frac{1}{2}\log(\det(-J)) - \frac{1}{2}\mathrm{Tr}(JC) . \tag{5.58}$$

and leads to the best set of couplings. The $Lq \times Lq$ symmetric matrix $J$ with elements $J_{ia,jb} \equiv J_{ij}(a,b)$ is negative definite (it corresponds to minus the precision matrix $T$ introduced in Chapter 3) and $C$ is the $Lq \times Lq$ connected correlation matrix: $C_{ia,jb} = f_{ij}(a,b) - f_i(a) f_j(b)$. Minimisation of Eq. (5.58) with respect to $J$ gives

$$\langle y_{ia} y_{jb}\rangle = -(J)^{-1}_{ia,jb} = C_{ia,jb} \qquad \Leftrightarrow \qquad J_{ij}(a,b) = -(C)^{-1}_{ia,jb} . \tag{5.59}$$

One should note at this point that $\sum_a C_{ia,jb} = 0$ because of Eq. (5.54). This implies that the $L$ vectors of the form $v_{ia}^{(k)} = \delta_{ik}$ are all zero eigenvectors of $C$, which is therefore not invertible. This is a consequence of the overparametrisation of the model. To address this issue, one may fix a gauge and thus reduce the dimension of $C$, changing $Lq \times Lq \to L(q-1) \times L(q-1)$ and thus suppressing the $L$ zero modes. One easy way is to eliminate one of the $q$ symbols when we introduce occurrence variables for the computation of $C$. In fact, the one-hot encoding with $q$ symbols is redundant because we know that exactly one amino acid (or gap) should be present on each site; hence, if all the first $q-1$ symbols are zero, the last symbol is necessarily one, and if at least one of the first $q-1$ is one, the last one should be zero. We can then perform a one-hot encoding using only the first $q-1$ symbols, hence obtaining a matrix of size $L(q-1)$, which is then invertible.

- **Undersampling and regularisation** - Once written in terms of the occurrence variables, the matrix $C$ can be expressed as

$$C_{ia,jb} = \langle y_{ia} y_{jb} \rangle = \frac{1}{M} \sum_{m=1}^{M} y_{ia}^m y_{jb}^m , \qquad \Rightarrow \qquad C = \frac{1}{M} Y^T Y , \qquad (5.60)$$

in terms of a $M \times L(q-1)$ matrix $Y_{m,ia} = y_{ia}^m$, whose rows are the sequences, in the occurrence variable representation. Undersampling, for example if $M < L(q-1)$ or if the sequences in the alignment are strongly correlated, can lead to additional zero modes in the correlation matrix $C$, which prevent its inversion needed for Eq. (5.59). In this tutorial we will use two methods to regularise the inverse problem: add a pseudo-count to the empirical frequencies and correlations, and add a $L_1$-norm regularisation.

- **Pseudo-count** - One can directly add a pseudo-count to the empirical frequencies and correlations in the following way:

$$f_i(a) \to f_i(a)(1-\alpha) + \frac{\alpha}{q} ,$$

$$(5.61)$$

$$f_{ij}(a,b) \to f_{ij}(a,b)(1-\alpha) + \frac{\alpha}{q^2} , \qquad \text{for } i \neq j .$$

Note that this transformation does not preserve the relation $f_{ii}(a,b) = f_i(a)\delta_{a,b}$, which must then be imposed by construction on the diagonal blocks. Such a pseudo-count gives a uniform prior probability if an amino acid $a$ on a site $i$ or a pair $(a,b)$ of amino acids on a pair of sites $(i,j)$ are never observed. Note that independently of the value of $\alpha$, the frequencies are still normalised to one. The regularisation parameter $\alpha$ should go to zero as the number of data increases, *i.e.* it should be of the order of $1/M$, as discussed in Chapter 4, but for contact prediction from a protein MSA its optimal value has been empirically found to be much larger, of the order of $\alpha = 0.5$ [38]. Adding such a pseudo-count is similar to the shrinkage method discussed in Chapter 4.

- $L_1$-**norm regularisation and graphical lasso** – Another possible regularisation scheme in the Gaussian approximation is to enforce sparsity of the coupling matrix through a $L_1$-norm prior. One has then to minimise, see section 5.1.2:

$$S_c^{L1} = -\frac{1}{2} \log \det (-J) - \frac{1}{2} \text{Tr}(J C) + \frac{\gamma}{2} \sum_{i,j,a,b} |J_{ij}(a,b)| . \qquad (5.62)$$

As before, $C$ is the $L(q-1) \times L(q-1)$ empirical covariance matrix with the last symbol removed, and $J$ is the negative definite $L(q-1) \times L(q-1)$ coupling matrix, satisfying the gradient equation

$$-J^{-1} - C + \gamma \operatorname{sign}(J) = 0 , \tag{5.63}$$

where the sign operation is applied element-wise to $J$. Let us define $W = -J^{-1}$, the model estimation of the covariance matrix. We can rewrite the matrices in two blocks, separating one variable (denoted by $i = 0$) from the others (denoted by $j$):

$$W = \begin{pmatrix} W_{00} & W_{0j} \\ W_{j0} & W^r \end{pmatrix} , \qquad C = \begin{pmatrix} C_{00} & C_{0j} \\ C_{j0} & C^r \end{pmatrix} . \tag{5.64}$$

Notice that $W^r$ is the matrix with line $i$ and column $i$ (denoted by 0 hereafter) removed, $W_{0j}$ is the corresponding column vector, and $W_{00}$ the diagonal element. All the values of $i$, comprised between 1 and $L$ are considered, one after the other. The definition of the inverse matrix implies

$$-\begin{pmatrix} W_{00} & W_{0j} \\ W_{j0} & W^r \end{pmatrix} \begin{pmatrix} J_{00} & J_{0j} \\ J_{j0} & J^r \end{pmatrix} = \begin{pmatrix} 1 & 0 \\ 0 & I \end{pmatrix} \quad \Rightarrow \quad -J_{00} W_{j0} - W^r J_{j0} = 0 , \tag{5.65}$$

which gives $W_{j0} = -W^r J_{0j}/J_{00} = W^r \beta$, where $\beta = -\frac{J_{j0}}{J_{00}}$. The lower block of Eq. (5.63) gives

$$W_{j0} - C_{j0} + \gamma \operatorname{sign}(J_{j0}) = 0 \quad \Rightarrow \quad W^r \beta - C_{j0} + \gamma \operatorname{sign}(\beta) = 0 , \tag{5.66}$$

recalling that $J_{00} < 0$ because $J$ is negative definite, and then $\operatorname{sign}(J_{j0}) = \operatorname{sign}(\beta)$. Eq. (5.66) shows that the vector $\beta$ can be estimated from the lasso algorithm we have studied in tutorial 4. Furthermore, because the diagonal elements $J_{00}$ are all negative, the diagonal terms in Eq. (5.63) give the condition $W_{00} - C_{00} - \gamma = 0$, or $W_{00} = C_{00} + \gamma$. This suggests the following algorithm [24, 39]:

---

### Graphical lasso algorithm

* Start from a matrix $W = C + \gamma I$; the diagonal elements are already correct and they are not updated anymore.
* Choose iteratively one line and column of $W$ and decompose the matrices in blocks. Compute $\beta$ using the lasso algorithm to solve Eq. (5.66). Update the corresponding line and column of $W$ by replacing $W_{j0} \leftarrow W^r \beta$.
* Repeat until convergence. After convergence, calculate $J = -W^{-1}$.

---

## 5.4.2   Problem

**Data:**
Data files can be downloaded from the book webpage[7], see tutorial 5 repository. The data file *seqPF14.txt* contains $M$=2143 aligned sequences of the trypsin inhibitor protein PF00014, each containing $L$=53 residues. The data have been taken from the PFAM database[8] (note that the database keeps evolving and the number of available sequences is now much larger). The structural information about the protein, obtained from the Protein Data Bank (PDB) database[9], is in the data file *backmap_distances_e_PF00014.txt*. The file contains $L(L-1)/2 = 1378$ rows. In each row, the first and second numbers are the two residue indices $(i < j)$, and the fifth number is their distance in Angstroms; the other numbers are irrelevant.

**Questions:**

1. Write the equations that give the model frequencies and pairwise correlations as averages over the Potts distributions given by Eq. (5.55). Recall how the minimisation of the cross entropy leads to the condition that the model and empirical frequencies coincide, and how this corresponds to the maximal entropy principle.

2. Read the data and convert the MSA alphabet from letters to numbers in the range $0, \ldots, 20$: use the routine that is given in the Jupyter notebook *td5_start.ipynb*, which provides the amino-acid alphabet.

3. Expand the MSA by replacing, on each line, the $L$ numbers $a_1, \ldots, a_L$ by the $L(q-1)$ one-hot encoded variables $s_{ia} \in \{0,1\}$, removing the last symbol to fix the gauge. One obtains in this way a binary matrix representation $\mathcal{S} = \{s_{ia}^m\}$ of the MSA of size $L(q-1) \times M$. The notebook *td5_start.ipynb* already contains this routine.

4. Extract from the binary representation $\mathcal{S}$ of the MSA the frequencies and correlations using

$$f_i(a) = \frac{1}{M} \sum_{m=1}^{M} s_{ia}^m, \qquad f_{ij}(a,b) = \frac{1}{M} \sum_{m=1}^{M} s_{ia}^m s_{jb}^m, \qquad (5.67)$$

and compute the correlation matrix $C$.

5. Add a pseudo-count to the frequencies as in Eq. (5.61), compute again the correlation matrix and invert it to obtain the coupling matrix $J$ using Eq. (5.59).

6. One way to find the strongest interacting residues is to consider the Frobenius norm of the $q \times q$ coupling matrix

$$\mathcal{F}_{ij} = \sum_{a,b} J_{ij}(a,b)^2 . \qquad (5.68)$$

This contact estimator has the drawback that it is gauge dependent. It has been empirically observed that the best performances are obtained by using coupling

[7]https://github.com/StatPhys2DataDrivenModel/DDM_Book_Tutorials
[8]PFAM PF00014, https://pfam.xfam.org/family/PF00014
[9]PDB 5PTI, https://www.rcsb.org/structure/5PTI

matrices in the zero-sum gauge where $\sum_a J_{ij}(a,b) = \sum_b J_{ij}(a,b) = 0$. In such gauge, which minimises $\mathcal{F}_{ij}$, coupling matrices are centred around zero, even if they can have large positive and negative entries. Transform the couplings in the zero-sum gauge and compute $\mathcal{F}_{ij}$ for each pair. The routine to obtain the Frobenius norm is given in *td5_start.ipynb*.

7. Rank the pairs according to their values of $\mathcal{F}_{ij}$. The pairs with higher $\mathcal{F}_{ij}$ are the candidates for being contacts. Read the distance between pairs $i < j$ from the structure data file, and consider as true contacts all pairs $i < j$ such that $d_{ij} < 8$ Angstroms. Plot the positive predictive value (PPV) for the contact prediction. Repeat the same calculation by restricting only to pairs that are far away in the protein sequence, such that $|i - j| > 4$. The reason is that pairs that are close along the protein chain are obviously also close in space (see figure 5.3).

8. The so-called average product correction [40] has been shown to improve the contact predictions from the Frobenius norm. It consists in transforming the Frobenius norms in the following way:

$$\mathcal{F}_{ij}^{APC} = \mathcal{F}_{ij} - \frac{\mathcal{F}_{*j}\,\mathcal{F}_{i*}}{\mathcal{F}_{**}} \,, \tag{5.69}$$

where the $*$ symbol denotes averaging over an index, for example $\mathcal{F}_{i*} = \sum_j \mathcal{F}_{ij}/L$. Such correction is useful to decrease couplings involving sites that appear to be connected to a large number of neighbours. Because a site must have a limited number of neighbours in the three-dimensional space, having very interconnected sites is a problem that can be due to the sampling limitations and to the approximations involved in the inference method. Rank the pairs according to their values of $\mathcal{F}_{ij}^{APC}$ and repeat the study of point 7. The routine given in *td5_start.ipynb* also includes the APC correction.

9. Implement the graphical lasso algorithm described above. Infer the couplings $J$ using this algorithm. Perform the same analysis as in points 6,7,8 above. Compute the PPV including the APC correction. What is the value of the regularisation parameter $\gamma$ that you expect to be optimal? Try that value of $\gamma$ and some other values and compare the results. Compare with the pseudo-count results. Calculate the number of non zero elements in the coupling matrix.

10. *Optional:* Use the mutual information between the occurrences of two amino acids in sites $i$ and $j$ obtained directly from the MSA,

$$MI_{ij} = \sum_{a,b} f_{ij}(a,b) \log \frac{f_{ij}(a,b)}{f_i(a)\,f_j(b)} \,, \tag{5.70}$$

to rank the contacts, and plot the PPV including pseudo-count and APC corrections also in this case. Note that $MI_{ij}$ is zero if the amino acids on sites $i$ and $j$ are independent, which implies $f_{ij}(a,b) = f_i(a)\,f_j(b)$. Such direct approach was indeed the first one to use the observed covariations in the MSA as predictors of contacts [40]. Verify that interactions are better predictors of contacts than direct correlations. Why?

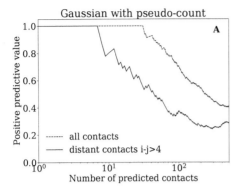
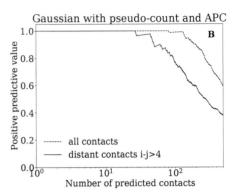

**Fig. 5.4** PPV for contact prediction, in the Gaussian approximation with pseudo-count $\alpha = 0.5$, without (**A**) and with (**B**) the APC correction.

### 5.4.3   Solution

The solution to the theoretical question 1 can easily be found in Chapter 5 and will not be reported here. The programme *tutorial5.ipynb* reads the MSA and computes the coupling matrix between the different amino acids on the different sites of the protein. It then uses this matrix to predict the contacts between residues in the three-dimensional structure of the protein. The solution to questions 2–6 (read the data, convert into numbers and then into the one-hot encoding, compute the frequencies, add a pseudo-count, invert the correlation matrix, compute the Frobenius norm) can be found in the programme. In the following, we discuss the results.

7. The positive predictive value (PPV) obtained by ranking the Frobenius norms of the couplings $\mathcal{F}_{ij}$ is given in figure 5.4**A** for all contacts and for contacts distant along the backbone $i - j > 4$. The first 30 couplings correspond to sites in contact and the first seven predictions for distant contacts are correct. Of the first 100 predictions for all couplings, about 70% are correct. Of the first 50 predictions for distant couplings, about 40% are correct

8. The PPV obtained by applying the APC correction to the $\mathcal{F}_{ij}$ is given in figure 5.4**B**. Predictions greatly improve with the APC: the first 80 couplings correspond to sites in contact and the first 28 predictions for distant contacts are correct. Of the first 100 predictions for all couplings, about 99% are correct. Of the first 50 predictions for distant couplings, about 80% are correct.

9. The programme uses the Gaussian approximation with a $L_1$-norm regularisation (graphical lasso or g-lasso algorithm) to derive the coupling matrix between the different amino acids on the different sites of the protein. The $\gamma$ parameter of the regularisation is estimated according to the large deviation theory of section 5.1.3,

$$\gamma \sim \sqrt{\log{(L \times q)}/M} = 0.057 \ . \tag{5.71}$$

This value is used in the following calculations. A comparison of the Frobenius norms of the couplings obtained from g-lasso with those obtained from the Gaussian approximation is shown in figure 5.5.

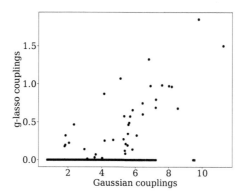

**Fig. 5.5** Comparison of the Frobenius norms of the couplings obtained within the Gaussian approximation with those obtained by the graphical lasso algorithm.

The PPV obtained by ranking the Frobenius norms of the g-lasso couplings is given in figure 5.6**A** for all contacts and for contacts distant along the backbone $i - j > 4$. The first 20 couplings correspond to sites in contact and the first four predictions for distant contacts are correct. Of the first 100 predictions for all couplings, about 40% are correct. Of the first 50 predictions for distant couplings, about 40% are correct

The PPV obtained by applying the APC to the g-lasso couplings is given in figure 5.6**B** for all contacts and for contacts distant along the backbone $i - j > 4$. Predictions slightly improve with the APC: the first 12 couplings correspond to sites in contact and the first four predictions for distant contacts are correct. Of the first 100 predictions for all couplings, about 40% are correct. Of the first 50 predictions for distant couplings, about 50% are correct. Performances are worse than the ones obtained by Gaussian inversion and pseudo-count.

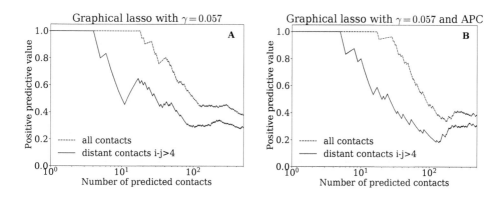

**Fig. 5.6** PPV for contact prediction, from the graphical lasso algorithm with $\gamma = 0.057$, without (**A**) and with (**B**) the APC correction.

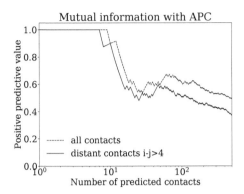

**Fig. 5.7** PPV for contact prediction, from the mutual information, with the APC correction.

10. The PPV obtained by ranking the mutual informations (MI) with APC, $MI_{ij}^{APC}$, is given in figure 5.7 for all contacts and for contacts distant along the backbone $i - j > 4$. Even with the APC corrections, MI predictions are poor especially for all contacts: the first nine couplings correspond to sites in contact and the first eight predictions for distant contacts are correct. Of the first 100 predictions for all couplings, about 60% are correct. Of the first 50 predictions for distant couplings, about 60% are correct.

# 6

# Unsupervised learning: from representations to generative models

Unsupervised learning is the task of inferring the statistical structure common to a set of data $Y = \{\boldsymbol{y}^m\}$, $m = 1, \cdots, M$. Unveiling this structure may be helpful for multiple purposes, such as interpreting the defining characteristics of the data, allowing for further processing, or even generating new data items. We have already encountered methods for unsupervised learning in the previous chapters. An example, discussed in Chapter 5, is that of Boltzmann machine learning of a probability distribution $p(\boldsymbol{y})$, which can then be sampled to produce new data. If the learning is good, the generated data should be statistically indistinguishable from the training data. For example, from a set of images of human faces, the machine can learn how to generate new "fake" images of faces. The principal component analysis (PCA) method discussed in Chapter 3 is one such example, in which one learns some features of the data (the directions in the data space along which data vary the most) that can be used to represent them.

The field of unsupervised learning is particularly vast. In this chapter we will only discuss a few examples that are connected to previous chapters, in particular autoencoders, restricted Boltzmann machines, and a revisited version of PCA. The emphasis will be put on two major goals of unsupervised learning, that is, the production of adequate internal representations of the data, and the generative character of the models.

## 6.1 Autoencoders

To illustrate the idea of internal representations we start with autoencoders. The idea is to build a machine capable of mapping $L$-dimensional inputs $\boldsymbol{y}$ onto themselves, with the constraint that the mapping has to go through a dimensional bottleneck, see figure 6.1**A**. The simplest architecture takes as input a data point $\boldsymbol{y}$ of dimension $L$, and computes an internal representation of dimension $K$:

$$z_\ell = \Phi\left(\sum_{i=1}^{L} J_{\ell,i}\, y_i\right) = \Phi\left(\boldsymbol{J}_\ell \cdot \boldsymbol{y}\right) , \qquad \ell = 1, \ldots, K .\tag{6.1}$$

The activation function $\Phi(I)$ is left generic for the moment. The internal units are then used to construct an output vector $\boldsymbol{x}$ of the same size as the input, as a non-linear combination of the internal units:

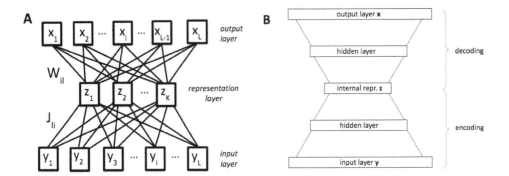

**Fig. 6.1** Sketches of autoencoders with one representation layer ($z$) directly connected to the input ($y$) and output ($x$) layers (**A**) and with intermediate hidden layers defining non-linear encoding and decoding operations (**B**).

$$x_i = \Psi \left( \sum_{\ell=1}^{K} W_{i,\ell}\, z_\ell \right) = \Psi\left(W_i \cdot z\right) , \qquad i = 1, \dots, L . \tag{6.2}$$

The autoencoder is then trained in such a way as to minimise the distance between the input and the output, *i.e.* $x \sim y$. In practice, this can be achieved by minimising a quadratic loss function

$$\mathcal{L}(\{J_\ell, W_\ell\}|Y) = \frac{1}{M} \sum_{m=1}^{M} [y^m - x(y^m)]^2 . \tag{6.3}$$

If $K \ll L$, this forces the autoencoder to learn efficient low-dimensional representations of the data, *i.e.* allowing for an accurate reconstruction the $L$-dimensional vectors $y$ from the $K$-dimensional internal unit vectors $z$.

### 6.1.1 Case of linear autoencoders

We will now show that PCA is a special case of an autoencoder. We will first consider the case $K = 1$ (one internal unit) and linear activation functions, $\Phi(I) = \Psi(I) = I$. Then, we have only one input ($J$) and output ($W$) weight vector. The output is given by

$$x = (J^T \cdot y)\, W . \tag{6.4}$$

To compare with the PCA discussed in Chapter 3, we assume that the input vectors $y$ have zero mean, $\langle y_i \rangle = 0$, and covariance matrix $\langle y_i y_j \rangle = \widehat{C}_{ij}$, with $\widehat{C}_{ii} = 1$. For an infinite amount of available data, $M \to \infty$, Eq. (6.3) becomes

$$\mathcal{L}(J, W) = \langle (y - (J^T \cdot y)\, W)^2 \rangle = \langle y^2 - 2(y^T \cdot W) \times (J^T \cdot y) + (J^T \cdot y)^2 W^2 \rangle$$
$$= L - 2\, J^T \cdot \widehat{C} \cdot W + W^2 \times J^T \cdot \widehat{C} \cdot J . \tag{6.5}$$

Hence, the equations for the weight vectors are obtained by

$$0 = \frac{\partial \mathcal{L}}{\partial \mathbf{J}} = -2\,\widehat{C} \cdot \mathbf{W} + 2\,\mathbf{W}^2 \times \widehat{C} \cdot \mathbf{J} \ ,$$

$$0 = \frac{\partial \mathcal{L}}{\partial \mathbf{W}} = -2\,\widehat{C} \cdot \mathbf{J} + 2\,(\mathbf{J}^T \cdot \widehat{C} \cdot \mathbf{J})\,\mathbf{W} \ . \tag{6.6}$$

Combining these two equations leads to

$$\widehat{C} \cdot \mathbf{J} = (\mathbf{J}^T \cdot \widehat{C} \cdot \mathbf{J})\mathbf{W} = \frac{1}{\mathbf{W}^2}\,\widehat{C} \cdot \mathbf{W} \qquad \Rightarrow \qquad \widehat{C} \cdot \mathbf{W} \propto \mathbf{W} \ , \tag{6.7}$$

which implies that $\mathbf{W}$ is proportional to one of the eigenvectors of $\widehat{C}$. Note that because $\mathbf{W}$ and $\mathbf{J}$ are multiplied in the output, see Eq. (6.4), we can fix the norm $|\mathbf{W}| = 1$ without loss of generality. Introducing the eigenvalues $\lambda^\alpha$ and the normalized eigenvectors $\mathbf{u}^\alpha$ of $\widehat{C}$, with $\alpha = 1, \ldots, L$, we have $\mathbf{W} = \mathbf{u}^\alpha$ for some $\alpha$. Next, from Eq. (6.6) above we get

$$\frac{\widehat{C} \cdot \mathbf{J}}{\mathbf{J}^T \cdot \widehat{C} \cdot \mathbf{J}} = \mathbf{W} = \mathbf{u}^\alpha \qquad \Rightarrow \qquad \mathbf{J} = \mathbf{u}^\alpha \ , \tag{6.8}$$

if all eigenvalues are strictly positive. Plugging the result $\mathbf{W} = \mathbf{J} = \mathbf{u}^\alpha$ into Eq. (6.5), we obtain

$$\mathcal{L} = L - \lambda^\alpha \ , \tag{6.9}$$

which implies that the loss is minimised by choosing the index $\alpha$ to correspond to the largest eigenvalue of $\widehat{C}$, *i.e.* to the principal component[1].

We conclude that a linear autoencoder, with a single internal unit, will learn the principal component of the data. This result can be easily generalised to the case of $K$ internal units (under the same assumptions), for which the autoencoder learns the $K$ principal components of $\widehat{C}$. The output of the autoencoder is then

$$\mathbf{x} = \sum_{\alpha=1}^{K} \left((\mathbf{u}^\alpha)^T \cdot \mathbf{y}\right) \mathbf{u}^\alpha \ , \tag{6.10}$$

which amounts to project $\mathbf{y}$ on the subspace spanned by its $K$ principal components. Given the low dimensionality of its internal representation, this is the choice that maximises the total variance of $\mathbf{x}$ and thus allows the machine to reproduce the input as much as possible.

### 6.1.2 Application to denoising

An important application of autoencoders is denoising. Assume that the input vectors $\mathbf{y}$ are corrupted by some noise $\boldsymbol{\eta}$, and we would like to recover the original vectors on the output layer $\mathbf{x}$ rather than simply reproducing them as before. Let us assume that the noise is uncorrelated from $\mathbf{y}$, with $\langle \eta_i \rangle = 0$ and $\langle \eta_i \eta_j \rangle = \eta^2 \delta_{ij}$. Suppose that we are given a set of $M$ examples of successful denoising[2], *i.e.* we have the noisy

---

[1]Note that $\operatorname{Tr} \widehat{C} = L$ and $\widehat{C}$ is non-negative, so $0 \leq \lambda^\alpha \leq L$, which guarantees that the loss is positive.

[2]This is obviously not the case in practice, but this assumption will allow us to understand how to modify the autoencoder to denoise the data.

vectors $y^m + \eta^m$ and the original vectors $y^m$. We can then train an autoencoder to perform the denoising, *i.e.* to associate to each input $y = y^m + \eta^m$ (the noisy vector) its corresponding output $x = y^m$ (the reconstructed vector). We consider the same setting as in section 6.1.1, using the loss function in Eq. (6.3) with $M \to \infty$, input vectors with $\langle \widehat{y}_i \rangle = 0$ and $\langle \widehat{y}_i \widehat{y}_j \rangle = \widehat{C}_{ij}$, and a simple linear autoencoder with a single internal unit. Because the output of the autoencoder is $x(y) = (J^T \cdot y)\, W$, the loss function reads

$$\mathcal{L}(\{J_\ell, W_\ell\}|Y) = \frac{1}{M} \sum_{m=1}^{M} \left[ y^m - x(y^m + \eta^m) \right]^2 , \tag{6.11}$$

which in the limit $M \to \infty$ tends to a simple modification of Eq. (6.5):

$$\mathcal{L}(J, W) = L - 2\, J^T \cdot \widehat{C} \cdot W + W^2 \times J^T \cdot (\widehat{C} + \eta^2 \mathcal{I}_L) \cdot J , \tag{6.12}$$

where $\mathcal{I}_L$ is the $L \times L$ identity matrix. The minimisation procedure of section 6.1.1 can be repeated, with the result

$$W = u^\alpha , \qquad J = \frac{\lambda^\alpha}{\lambda^\alpha + \eta^2}\, u^\alpha , \tag{6.13}$$

where $u^\alpha$ is the principal component of $\widehat{C}$ and $\lambda^\alpha$ its associated eigenvalue. As in the previous case the autoencoder learns to project the data on the direction that maximises the variance. However, Eq. (6.13) shows that the weights $J$ associated to eigenvalues $\lambda^\alpha$ much smaller than $\eta^2$ essentially vanish: the level of noise in the data implicitly fixes the effective dimension of representations to the number of eigenvalues larger than $\eta^2$.

Of course, denoising can be (and should be) performed with more complex architectures, such as those described in section 6.1.3, featuring multiple internal units, hidden layers, and non-linear activation functions. In that case the loss function must be minimised numerically and one can achieve good performance at denoising (*e.g.* images) in practical applications.

### 6.1.3 Sparse representations

While our linear autoencoder rightly identifies the best low-dimensional representation, the quality of reconstruction of the inputs is poor: unless $\widehat{C}$ is of low rank, the loss $\mathcal{L}$ is not small. How could we find representations that extract meaningful low-dimensional features of the data and, yet, allow for accurate reconstruction?

One possibility is offered by sparsity. Consider again the autoencoder in figure 6.1**A**. Let us relax the constraint that the representation layer should be of low dimension, and let $K$ take any value, even larger than $L$. However, let us demand that, for any input $y$, many entries $z_\ell$ of the representation vector vanish. A practical way to impose sparsity in the representation is through the addition of a $L_1$-norm regularisation, which amounts to modify the loss in Eq. (6.3) into

$$\mathcal{L}(J, W|Y) = \frac{1}{M} \sum_{m=1}^{M} \left[ y^m - x(y^m) \right]^2 + \frac{\gamma}{M} \sum_{m=1}^{M} ||z(y^m)||_1 . \tag{6.14}$$

With an appropriate choice of the penalty strength $\gamma$ we can make sure that, on average over the inputs, only $K' \ll L$ entries of the representations are not equal to zero. Hence, as in the previous formulation of the autoencoder, the output is built from a very limited number of $W_i$ feature vectors, see Eq. (6.2). However, the repertoire of available features is now much larger, which allows the machine to select the few, most adequate ones for each input. Informally speaking, the situation is similar to what we do when expressing a sentence. Though the English dictionary contains hundreds of thousands of words, we pick up and assemble only a few of them to make a sentence. The combinatorial variety of choices of words results in huge expressivity. In the case of linear autoencoders, minimising the loss in Eq. (6.14) will result in a trade-off between extracting the top components of the data and making these components sparse. We will come back to the concept of sparse PCA in the context of online learning, see section 6.4.2.

Let us pursue this analogy with words and dictionaries. The loss in Eq. (6.14) relies on the two essential parts of the autoencoder architecture in figure 6.1:

- the encoder, which computes the representation $z$ from the input $y$;
- the decoder, which computes the output $x$ from the representation $z$.

The most important constraint, as far as sparsity is concerned, affects the decoding step: each input should be expressed as a combination of a small number of feature vectors. How or if the non-zero coefficients $z_k$ in this combination are computed from the input (the encoding step) is somewhat irrelevant here. We therefore discard the encoder part, and turn the previous loss into

$$\mathcal{L}(z, W|Y) = \frac{1}{M} \sum_{m=1}^{M} \sum_{i=1}^{L} \left[ y_i^m - \Psi(W_i \cdot z^m) \right]^2 + \frac{\gamma}{M} \sum_{m=1}^{M} ||z^m||_1 , \quad (6.15)$$

which has to be minimised over the representations $z^m$, as well as over the decoder weights $W$. In words, we are looking for a low-dimensional representation of our input data $y^m$ based on the "dictionary" of features $W_1, \ldots, W_K$ and on coefficients $z^m$ forced to be sparse through the $L_1$-norm penalty (or any other penalty). The search of $W$ minimising Eq. (6.15) is called dictionary learning.

### 6.1.4 Multilayer and non-linear autoencoders

Another way to improve over linear autoencoders and PCA is to implement more complex input-output mappings. This can be done in two complementary ways.

First, keeping the same architecture as in figure 6.1**A**, we can consider non-linear activation functions $\Phi(I)$ and $\Psi(I)$ in Eqs. (6.1) and (6.2), such as the so-called rectified linear units (ReLU) defined as $\max(a\,I - b, 0)$. The defining parameters of the activation functions ($a, b$ in the ReLU case) can be determined through the minimisation of the loss.

Second, one can also consider multilayer autoencoders, see figure 6.1**B**. Using the compact notation $z = \Phi(J^T \cdot y)$ for Eq. (6.1), in which it is implicitly understood that the weight matrix $J = \{J_\ell\}$ is first applied to the input $y$ and then the activation function $\Phi(I)$ is computed component-wise, we can write the "encoding" part of a multilayer autoencoder as

$$z = \Phi_E\big(J_E^T \cdot \Phi_{E-1}\big(J_{E-1}^T \cdots \Phi_2\big(J_2^T \cdot \Phi_1\big(J_1^T \cdot y\big)\big)\big)\big) \,, \tag{6.16}$$

where $z$ is again the internal representation of the input, of dimension $K$, here obtained via $E$ encoding layers with activation functions $\Phi_e$ and weight matrices $J_e$. Notice that, as the depth $E$ of the encoder increases, the non-linearities progressively transform any arbitrary distribution over the inputs $y$ into a Gaussian-like distribution over the representations $z$. We will come back to this point later.

The internal representation is then "decoded" according to an extension of Eq. (6.2) to the multilayer situation:

$$x = \Psi_D\big(W_D \cdot \Psi_{D-1}\big(W_{D-1} \cdots \Psi_2\big(W_2 \cdot \Psi_1\big(W_1 \cdot z\big)\big)\big)\big) \,, \tag{6.17}$$

here with $D$ decoding layers with activation functions $\Psi_d(I)$ and weight matrices $W_d$. Generically, we want the internal representation to be of lower dimension of the input, $K \ll L$, and it is then reasonable to decrease progressively the size of the hidden layer during the encoding, and to increase it progressively during the decoding, as illustrated in figure 6.1**B**.

As in the previous cases, we can minimise the loss function in Eq. (6.3), now a function of all weight matrices, to train the autoencoder on a given dataset. The minimisation has now to be performed numerically, for example via gradient descent algorithms. $L_1$-norm or $L_2$-norm penalties on each weight matrices can be considered, as well as specific architectures in which the weight matrices are not dense, but have non-zero elements only in some particular positions (*e.g.* so-called convolutional layers). Note that the quadratic loss function can also be replaced by other positive functions, depending on the nature of the input vectors.

## 6.2 Restricted Boltzmann machines and representations

As we have seen in Chapters 3 and 5, multivariate Gaussian distributions can be either described in terms of a graph of dependencies between variables or of special directions (the top components of the covariance matrix) in the data space along which large fluctuations take place. The coordinates along these top components offer a low-dimensional relevant representation of the data, at the core of principal component analysis.

The framework of Boltzmann machines (BM) naturally extends the notion of graph of dependencies $T$ into that of network of interactions $J$. The goal of this section is to show that the notions of directions or representations are also fruitful for these machines, through the introduction of a new architecture, the so-called restricted Boltzmann machines.

### 6.2.1 Alternative formulation of Boltzmann machines

Consider the BM over $\sigma_i = 0, 1$ variables defined in Eqs. (5.24) and (5.28). Its probability distribution can be written in the form,

$$p(\boldsymbol{\sigma}) = \frac{1}{Z} \exp\left(\sum_i g_i \sigma_i + \frac{1}{2} \sum_{i,j} J_{ij} \sigma_i \sigma_j\right) \,, \tag{6.18}$$

where we changed the notation for the fields, now called $g_i$, and the couplings on the diagonal, $J_{ii}$, vanish. The value of the diagonal couplings is actually arbitrary: setting $J_{ii} = J_0 \neq 0$ can be easily compensated through a redefinition of the fields, $g_i \rightarrow g_i - \frac{1}{2}J_0$, since $\sigma_i^2 = \sigma_i$. Without loss of generality, we can therefore assume that all the eigenvalues $\lambda_\mu$ of the coupling matrix are strictly positive (by choosing $J_0$ large enough). We therefore write

$$J_{ij} = \sum_{\mu=1}^{L} \lambda_\mu \, v_{\mu i} \, v_{\mu j} \ , \tag{6.19}$$

where the $v_\mu$'s are the normalised eigenvectors of $J$. Plugging this expression into Eq. (6.18) we obtain

$$p(\boldsymbol{\sigma}) = \frac{1}{Z} \exp\left( \sum_i g_i \, \sigma_i + \frac{1}{2} \sum_{\mu=1}^{L} \lambda_\mu \left( \sum_i v_{\mu i} \, \sigma_i \right)^2 \right) . \tag{6.20}$$

Using the following identity, valid for any real-valued $x$,

$$e^{x^2/2} = \int \frac{dh}{\sqrt{2\pi}} \, e^{-h^2/2 + hx} \ , \tag{6.21}$$

the square term in Eq. (6.20) can be written as an $L$-dimensional Gaussian integral over an auxiliary vector $\boldsymbol{h}$,

$$p(\boldsymbol{\sigma}) = \frac{1}{Z} \int d\boldsymbol{h} \, \exp\left( \sum_i g_i \, \sigma_i - \frac{1}{2} \sum_\mu h_\mu^2 + \sum_{\mu,i} h_\mu \, w_{\mu i} \, \sigma_i \right) , \tag{6.22}$$

where $w_{\mu i} = \sqrt{\lambda_\mu} \, v_{\mu i}$ and the factor $(2\pi)^{L/2}$ has been absorbed in a redefinition of $Z$. Eq. (6.22) suggests to interpret the probability over $\boldsymbol{\sigma}$ as the marginal of a joint distribution over both $\boldsymbol{\sigma}$ and $\boldsymbol{h}$,

$$p(\boldsymbol{\sigma}) = \int d\boldsymbol{h} \, p(\boldsymbol{\sigma}, \boldsymbol{h}) \ , \quad \text{where} \quad p(\boldsymbol{\sigma}, \boldsymbol{h}) = \frac{1}{Z} \exp\left( - E(\boldsymbol{\sigma}, \boldsymbol{h}) \right) \tag{6.23}$$

is an exponential (Boltzmann) distribution with associated energy

$$E(\boldsymbol{\sigma}, \boldsymbol{h}) = \sum_\mu U_\mu(h_\mu) - \sum_i g_i \, \sigma_i - \sum_{\mu,i} h_\mu w_{\mu i} \, \sigma_i \ , \tag{6.24}$$

and the potential acting on the hidden variables is quadratic:

$$U_\mu(h) = \frac{h^2}{2} \ . \tag{6.25}$$

The energy function in Eq. (6.24) defines a model, whose graph of dependencies is bipartite (figure 6.2): each variable $\sigma_i$ is connected to the variables $h_\mu$, and, vice versa,

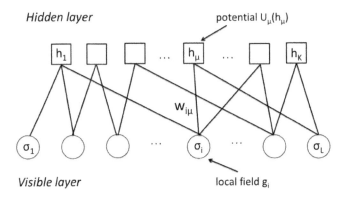

**Fig. 6.2** Architecture of restricted Boltzmann machines. The set of interactions $w_{i\mu}$ define a bipartite graph, connecting $L$ visible units $\sigma_i$ and $K$ hidden units $h_\mu$, according to the energy in Eq. (6.24). In addition, local fields $g_i$ act on the units in the visible layer, and potentials $U_\mu(h_\mu)$ on the units in the hidden layer.

$h_\mu$ is connected to the variables $\sigma_i$. No interaction can exist between any two $\sigma$'s or any two $h$'s. For this reason, this model is called restricted Boltzmann machine (RBM). Let us remark that the above derivation of the energy function is not specific to binary variables, and could easily be extended to the case of Boltzmann machines defined on categorical variables, see Eq. (5.25), with the result

$$E(\boldsymbol{\sigma}, \boldsymbol{h}) = \sum_\mu U_\mu(h_\mu) - \sum_i g_i(\sigma_i) - \sum_{\mu,i} h_\mu w_{\mu i}(\sigma_i) \ . \tag{6.26}$$

We concentrate on binary variables in the following for the sake of simplicity.

The $\boldsymbol{h}$ variables are called hidden variables, because there is no direct access to their values from the data; while they were introduced through the Gaussian trick of Eq. (6.21), they actually have a deep meaning, as we will discuss in the next section. In contradistinction, the original $\boldsymbol{\sigma}$ variables are referred to as visible variables. Notice that the number of hidden units, $K$, can be smaller than the number of visible units, $L$, if the interaction matrix $J$ has zero eigenvalues, see Eq. (6.19). We will see below more general expressions of the energy function $E$, in which $K$ can be larger than $L$.

The joint distribution $p(\boldsymbol{\sigma}, \boldsymbol{h})$ allows us to define the conditional probability of a hidden configuration given the visible configuration,

$$p(\boldsymbol{h}|\boldsymbol{\sigma}) = \prod_{\mu=1}^K p_\mu(h_\mu|\boldsymbol{\sigma}) \ , \quad \text{where} \quad p_\mu(h_\mu|\boldsymbol{\sigma}) = \frac{e^{-U_\mu(h_\mu)+h_\mu I_\mu(\boldsymbol{\sigma})}}{\int dh \ e^{-U_\mu(h)+h I_\mu(\boldsymbol{\sigma})}} \ , \tag{6.27}$$

and

$$I_\mu(\boldsymbol{\sigma}) = \sum_{i=1}^L w_{\mu i} \sigma_i \ . \tag{6.28}$$

The above formulae have immediate interpretations. When the visible layer carries the configuration $\boldsymbol{\sigma}$, each hidden unit $\mu$ receives the input $I_\mu(\boldsymbol{\sigma})$ defined in Eq. (6.28).

According to the expression of the potential in Eq. (6.25), the conditional probability of $h_\mu$ is therefore Gaussian, with unit variance and mean equal to the input $I_\mu(\boldsymbol{\sigma})$. Due to the bipartite nature of the interaction graph in figure 6.2, the conditional probability of $\boldsymbol{h}$ factorises over the hidden units, see Eq. (6.27). This property is useful, as it allows for an easy sampling of the conditional probability $p(\boldsymbol{h}|\boldsymbol{\sigma})$.

Assume now that we have at our disposal a configuration $\boldsymbol{h}$ of the hidden units. We can immediately write the conditional probability of a visible configuration $\boldsymbol{\sigma}$ as

$$p(\boldsymbol{\sigma}|\boldsymbol{h}) = \prod_{i=1}^{L} p_i(\sigma_i|\boldsymbol{h}) , \quad \text{where} \quad p_i(\sigma_i|\boldsymbol{h}) = \frac{e^{(g_i + I_i(\boldsymbol{h}))\,\sigma_i}}{1 + e^{g_i + I_i(\boldsymbol{h})}} , \tag{6.29}$$

and

$$I_i(\boldsymbol{h}) = \sum_{\mu=1}^{K} w_{\mu i}\, h_\mu . \tag{6.30}$$

Each visible variable is subject, in addition to its local field $g_i$, to an input $I_i(\boldsymbol{h})$ coming from the hidden layer. Again, this conditional probability is factorised, this time over the sites of the visible layer, which makes the sampling of $\boldsymbol{\sigma}$ easy from a practical point of view.

Alternate sampling from Eqs. (6.27) and (6.29) permits one to generate time sequences of visible and hidden configurations. This back-and-forth procedure offers a practical way, for long enough sequences, to draw visible configurations $\boldsymbol{\sigma}$ from the marginal distribution $p(\boldsymbol{\sigma})$.

### 6.2.2 Interplay between visible and hidden variables

What is the meaning of the hidden configurations $\boldsymbol{h}$? To get some intuition let us consider the simple case of a RBM with one hidden unit connected to a large visible layer ($L \gg 1$) through a set of interactions $w_i$ (where we have dropped the index $\mu = 1$). Half of these couplings are equal to $w_i = +2w/\sqrt{L}$, and the remaining half to $-2w/\sqrt{L}$, where $w > 0$ is fixed. Furthermore we set all local fields to $g_i = 0$. Let us start from a value $h = x\sqrt{L} > 0$ of the hidden unit, and sample the visible configuration from Eq. (6.29). What is the distribution of the inputs $I(\boldsymbol{\sigma}) = \sum_i w_i \sigma_i$ in Eq. (6.28) acting in turn on the hidden unit? As all the visible units are independent (for a fixed $h$ or $x$), the central limit theorem applies for large $L$, and we conclude that $I(\boldsymbol{\sigma})$ is a Gaussian random variable with mean

$$\langle I \rangle(h) = \sum_i \frac{w_i\, e^{h\,w_i}}{1 + e^{h\,w_i}} = \sqrt{L}\, w\, \tanh(x\,w) , \tag{6.31}$$

and variance

$$\mathrm{Var}[I](h) = \sum_i \frac{w_i^2\, e^{h\,w_i}}{(1 + e^{h\,w_i})^2} = \frac{w^2}{\cosh^2(x\,w)} . \tag{6.32}$$

If we now sample a new value of the hidden unit, say, $h' = x'\sqrt{L}$, from Eq. (6.27), then $x'$ will be, up to negligible fluctuations when $L \gg 1$, equal to $\langle I \rangle(h)/\sqrt{L}$. Alternate sampling of visible and hidden configurations is therefore equivalent to iterations of the mapping $x \to x' = w \tanh(x\,w)$. Eventually the value of the hidden unit will settle in the fixed point of this mapping. Two cases are possible:

- if $|w| > w_c = 1$, there is an attractive fixed point $x^* > 0$, with $x^* \to |w|$ for $|w| \gg w_c$. The value of the hidden unit will converge towards $x^* \sqrt{L}$, and the mean value of visible unit $i$ to

$$\langle \sigma_i \rangle = \frac{1}{2} \left[ 1 + \text{sign}(w_i) \tanh(x^* w) \right] , \qquad (6.33)$$

  according to Eq. (6.29).
- if $|w| < w_c$, the attractive fixed point is $x^* = 0$. Hence the hidden unit activation does not scale as $\sqrt{L}$ any longer, and shows Gaussian fluctuations with zero mean and variance of the order of 1 (compared to $L$). As a result, after many sampling steps, the mean value of $\sigma_i$ reaches $\frac{1}{2}$.

What have we learned from this toy model? If the couplings are too small, then the presence of the hidden unit has vanishing influence on the visible configurations, and our RBM is equivalent to an independent-site model. Conversely, for large enough couplings, a strong interdependence appears between the visible configurations $\boldsymbol{\sigma}$ and the hidden units $h$: $\boldsymbol{\sigma}$ is aligned along the weight vector $\boldsymbol{w}$ (visible units are more likely to take value 1 or 0 if their incoming connections are, respectively, positive or negative), and sends a very large input to the hidden unit. In turn, the strongly activated hidden unit stabilises the visible configuration, and so on.

We may now leave this toy model and go back to the general case of a RBM with $K$ hidden units. We expect large $h_\mu$ to exert strong influences on the visible configurations $\boldsymbol{\sigma}$, more precisely, to make them similar to their corresponding interaction vectors $\boldsymbol{w}_\mu = \{w_{\mu i}, i = 1, \ldots, L\}$. The larger $h_\mu$, the more similar $\boldsymbol{\sigma}$ will be to $\boldsymbol{w}_\mu$. In turn, given a visible configuration $\boldsymbol{\sigma}$, the hidden configuration $\boldsymbol{h}$ will inform us about the similarities between $\boldsymbol{\sigma}$ and the directions of interest $\boldsymbol{w}_\mu$. We can therefore think of the hidden units as coordinates in an adequate space of representations for the visible configurations, in much the same way as the projections of Gaussian multi-dimensional variables along the top components identified by PCA in Chapter 3.

### 6.2.3 Non-linear activation functions

While we have seen that rewriting a BM into a RBM naturally yields quadratic potentials over the hidden units, see Eq. (6.25), it is perfectly licit to consider RBM with non-quadratic potentials. Admissible potentials $U(h)$ should grow fast enough (more than linearly) for large $|h|$ so that the integral in Eq. (6.23) converges and the marginal distribution of $\boldsymbol{\sigma}$ is well defined.

To understand the effect of $U$ let us consider the average value of a hidden unit receiving an input $I$ from the visible layer:

$$\langle h \rangle (I) = \frac{\int \mathrm{d}h \, h \, e^{-U(h) + h\,I}}{\int \mathrm{d}h \, e^{-U(h) + h\,I}} . \qquad (6.34)$$

For large inputs $I$ this integral is dominated by the value of $h$ such that $U(h) - h\,I$ is minimal, and we conclude that

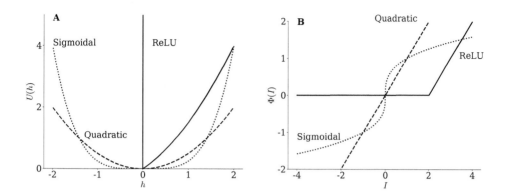

**Fig. 6.3 A.** Examples of potentials $U(h)$. ReLU (full line): $U(h) = +\infty$ for $h < 0$, $U(h) = \frac{h^2}{2} + I_c h$ for $h \geq 0$, where the threshold value is $I_c = 1$. Quadratic (dashed): $U(h) = \frac{h^2}{2}$. Sigmoidal (dotted): $U(h) = \frac{h^4}{4}$. **B.** Corresponding activation functions $\Phi(I)$.

$$\langle h \rangle(I) \simeq \Phi(I) , \quad \text{where} \quad \Phi = \left( U' \right)^{-1} \tag{6.35}$$

is the reciprocal function of the derivative of the potential and is called 'activation function'. In the case of the quadratic potential of Eq. (6.25) we have $\Phi(I) = I$: the activation function is linear, see figure 6.3**B**.

Non-quadratic potentials yield non-linear activation functions for large inputs. For instance, for $\alpha > 0$,

$$U(h) = \frac{\alpha}{\alpha + 1} |h|^{1+1/\alpha} \quad \longrightarrow \quad \Phi(I) = \text{sign}(I) \times |I|^\alpha . \tag{6.36}$$

Two regimes appear[3]:

- $\alpha > 1$: the activation function grows faster than linearly with $I$. In particular, for integer-valued $\alpha$, the contribution of this hidden unit to the log-likelihood is[4]

$$\Delta \log p(\boldsymbol{\sigma}) = \frac{1}{\alpha + 1} I(\boldsymbol{\sigma})^{\alpha+1} = \frac{1}{\alpha + 1} \sum_{i_1, i_2, \ldots, i_{\alpha+1}} w_{i_1} w_{i_2} \ldots w_{i_{\alpha+1}} \sigma_{i_1} \sigma_{i_2} \ldots \sigma_{i_{\alpha+1}} , \tag{6.37}$$

where the $w_i$'s denote its interactions with the visible units. If more than one hidden units, say, $\mu = 1, 2, \ldots, K_\alpha$ are subjected to potentials $U_\mu$ characterised by the same exponent $\alpha$, then their contributions to the log-likelihood sum up, see Eq. (6.37) with the corresponding weights $w_{\mu i}$. The effect onto the visible layer is to create $(\alpha + 1)$-body interactions terms between the visible units,

$$J_{i_1, i_2, \ldots, i_{\alpha+1}} \propto \sum_{\mu=1}^{K_\alpha} w_{\mu, i_1} w_{\mu, i_2} \ldots w_{\mu, i_{\alpha+1}} . \tag{6.38}$$

---

[3]Linear activation is found back for $\alpha = 1$.

[4]The dominant contribution to the log-likelihood in Eq. (6.23) is, for large inputs $I$, $\Delta \log p(\boldsymbol{\sigma}) \simeq -U(h) + h I(\boldsymbol{\sigma})$ where $h = \Phi(I(\boldsymbol{\sigma}))$. Eq. (6.36) then immediately leads to Eq. (6.37).

This formula is a generalisation of Eq. (6.19), and shows that RBM, contrary to BM, are not limited to pairwise interactions.

- $0 < \alpha < 1$: the activation function has a sigmoidal shape, see figure 6.3**B**, and grows slowly for large $|I|$. In the limit $\alpha \to 0^+$ the hidden unit $h$ can take values $+1$ or $-1$ only, depending on the sign of the input. Such hidden units are called Bernoulli. When the symmetry between the two values is lifted through the introduction of a field $u_\mu$ on hidden unit $\mu$, the probability of a visible configuration becomes

$$p(\boldsymbol{\sigma}) = \sum_{h_1=\pm 1} \sum_{h_2=\pm 1} \cdots \sum_{h_K=\pm 1} \exp\left( \sum_\mu u_\mu \, h_\mu + \sum_i g_i \, \sigma_i + \sum_{\mu,i} h_\mu \, w_{\mu i} \, \sigma_i \right) .$$

(6.39)

Hence, the distribution of $\boldsymbol{\sigma}$ is a mixture of $2^K$ independent-site models for visible units.

Let us emphasise that the behaviour of the activation function for small inputs $I$ is also very important. As we have seen in section 6.2.2, a large input $I_\mu$ signals a coherence between the visible configuration and the weight vector attached to hidden unit $\mu$. Even if $\boldsymbol{\sigma}$ and $\boldsymbol{w}_\mu$ are unrelated, their scalar product may take non-zero value, which in turn will modify $\boldsymbol{\sigma}$ at later steps of the sampling procedure. To avoid the accumulation of noise-induced effects, it is judicious to avoid small inputs onto hidden units. An efficient noise-suppression mechanism relies on the introduction of a threshold $I_c$ on the minimal value of inputs capable of triggering a non-zero activation of hidden units. An example is shown in figure 6.3**B**, and corresponds to

$$\Phi(I) = \max(0, I - I_c) .$$

(6.40)

This activation function is the rectified linear unit (ReLU) we have already encountered in section 6.1.4; the associated potential is $U(h) = \frac{h^2}{2} + I_c h$ if $h > 0$, and $U(h) = +\infty$ if $h < 0$ (figure 6.3**A**). In addition ReLU units are non-symmetric: the activity $h$ is always non-negative, and can be interpreted as the degree of presence of the feature $\boldsymbol{w}_\mu$ in the configuration $\boldsymbol{\sigma}$.

### 6.2.4   Learning RBM and features from data

While we have studied above the operation of RBM once their defining parameters[5] $\boldsymbol{\theta} = \{g_i, w_{\mu i}, U_\mu\}$ are specified, we have not discussed how these can be inferred from data yet. Assume we are given a set of visible configurations $\boldsymbol{\sigma}^m$, $m = 1, \ldots, M$. The log-likelihood per data point is given by

$$\mathcal{L}(\boldsymbol{\theta}) = \frac{1}{M} \sum_m \log p(\boldsymbol{\sigma}^m; \boldsymbol{\theta}) ,$$

(6.41)

---

[5]The potential $U$ is in principle a full function of its argument $h$, and requires an infinite number of parameters to be entirely described. In practice, however, one often chooses $U$ in a conveniently parametrised class of potentials, *e.g.* $U(h) = \frac{\gamma}{2}h^2 + I_c h$ for ReLU potentials on the $h > 0$ semi-line, which is spanned by a finite number of parameters, here, two.

where the probability of a visible configuration is given by Eq. (6.23) and we have made its dependence on $\boldsymbol{\theta}$ explicit in the notation. Let us write the gradient of $\mathcal{L}$ with respect to the components of $\boldsymbol{\theta}$, e.g. the weight $w_{\mu i}$. We obtain

$$\frac{\partial}{\partial w_{\mu i}} \mathcal{L}(\boldsymbol{\theta}) = \frac{1}{M} \sum_m \langle h_\mu \rangle (\boldsymbol{\sigma}^m) \, \sigma_i^m - \langle h_\mu \, \sigma_i \rangle \;, \tag{6.42}$$

where the conditional value of a hidden unit was defined in Eq. (6.34) and

$$\langle h_\mu \, \sigma_i \rangle = \int d\boldsymbol{h} \sum_{\boldsymbol{\sigma}} p(\boldsymbol{\sigma}, \boldsymbol{h}) \, h_\mu \, \sigma_i \;, \tag{6.43}$$

is the correlation between the values of the hidden and visible units according to the RBM. We recognize that the gradient vanishes when this model correlation matches its counterpart computed from the data[6].

To learn the RBM parameters from the data, we can follow a procedure similar to the Boltzmann machine learning described in section 5.2.3. The difficulty is to compute for each learning step the average values of observables over the model distribution, which requires back-and-forth Monte Carlo simulations or the use of approximate sampling such as contrastive divergence [28].

How are the weights evolving in the course of learning? Let us start from the solution of the independent-site problem: the local fields $g_i^0$ are such that the visible units reproduce the one-point statistics in the data, and the couplings $w_{\mu i}^0$ are null. Then, we learn the weights by ascending the gradient of $\mathcal{L}$, that is, with a learning dynamics of the type

$$w_{\mu i}^{t+1} = w_{\mu i}^t + \eta_w \frac{\partial \mathcal{L}}{\partial w_{\mu i}} \;, \tag{6.44}$$

where $t$ is the step of updating of the weights, and $\eta_w$ the learning rate. Obviously the other parameters of the RBM (the potentials $U_\mu$ and the fields $g_i$) have to be updated according to similar gradient-ascent equations. As the weights will remain very small for some (learning) time, we may linearize the gradient in Eq. (6.42), with the result

$$w_{\mu i}^{t+1} = w_{\mu i}^t + \eta_w \sum_{j(\neq i)} C_{ij} \, w_{\mu j}^t + O(w^2) \;, \tag{6.45}$$

where $C_{ij}$ are the entries of the empirical covariance matrix estimated from the data. We see that:

- $w_{i\mu} = 0$ is a fixed point of the dynamics. To avoid being stuck we may simply set a few of the $w_{\mu i}$ interactions to small, non-zero random values.
- The empirical covariance matrix, after removal of its diagonal elements, has zero trace. Hence, at least one of the eigenvalues is positive[7]. Eq. (6.45) tells us that the

---

[6]Strictly speaking, the latter quantity, $\frac{1}{M} \sum_m \langle h_\mu \rangle (\boldsymbol{\sigma}^m) \, \sigma_i^m$ also depends on the model parameters. Indeed, the value of hidden unit $\mu$ in the representation of the data point $\boldsymbol{\sigma}^m$ is not directly available and is estimated from the model through $\langle h_\mu \rangle (\boldsymbol{\sigma}^m)$.

[7]All eigenvalues vanish only if the visible units are uncorrelated in the data, *i.e.* $C_{ij} = 0$ for all $i \neq j$. In this case the weights remain zero as they should, because the independent-site model is exact.

weights will grow exponentially along the top component (associated to the largest eigenvalue). Hence, we see that the RBM correctly learns the most important feature of the data, as would PCA do.

Including higher-order terms in powers of $w$ in the evolution Eq. (6.45) for the weights has two effects. First, the weight vectors $\boldsymbol{w}_\mu$ will stop aligning along the same top component and start repel each other, to unveil more features of the data distribution. Second, for non-linear activation functions, the weight vectors will depart from the eigenvectors of $C$. Each hidden unit $\mu$ may develop large weights $w_{i\mu}$ on a subset of the visible sites $i$, whose size is related to the scale of correlations in the data [41]. In other words, hidden units acquire receptive fields and are active when specific features are present in the data. This property is reminiscent of the behaviour of neural cells in the visual cortex of mammals [42].

## 6.3  Generative models

An important application of unsupervised learning is generative modelling, which is defined as the following task. Given $M$ training data points $Y = \{\boldsymbol{y}^m\}_{m=1,...,M}$, one wishes to learn a probability distribution $p(\boldsymbol{y})$, such that samples generated from $p(\boldsymbol{y})$ are statistically indistinguishable from the training data. For example, the training data could be a set of "natural" images of human faces, and a proper generative model learned on them should be able to produce, via sampling of $p(\boldsymbol{y})$, new "artificial" images that could not be distinguished (in this case, by a human observer) from the natural ones. In this section we will give a few examples of generative models, revisiting some notions already introduced in Chapter 5, and will discuss one important application, namely, the design of functional proteins.

### 6.3.1  Boltzmann machines

A simple yet powerful class of generative models are Boltzmann machines, introduced in section 5.2. To follow the notations of that section, we consider here $q$-state categorical variables $\boldsymbol{\sigma} = (\sigma_1, \ldots, \sigma_L)$ and in particular the case $q = 2$ with $\sigma_i \in \{0,1\}$ as training data. We described in section 5.2.3 how a Boltzmann machine with local fields and pairwise interactions can be trained to define a distribution $p(\boldsymbol{\sigma})$, which reproduces the one- and two-point statistics of the training data, see Eq. (5.30). Furthermore, this Boltzmann machine is, consistently with the maximum entropy principle, see Eq. (5.23), the probability distribution that has maximal entropy among those which reproduce these low-order statistics.

We can now generate new "artificial" samples from $p(\boldsymbol{\sigma})$, *e.g.* via Monte Carlo sampling. How can we decide if our Boltzmann machine is a good generative model, *i.e.* if the sampled data are statistically indistinguishable from data?

A first criterion is that moments of order larger than or equal to three of test data, which have not been used during the Boltzmann machine learning procedure (see figure 4.2) should match their counterparts computed on model-generated data. For example, one can compare the three-point statistics of empirical data $\mu_{ijk}$ and that of the model:

$$\mu_{ijk} = \frac{1}{M} \sum_{m=1}^{M} \sigma_i^m \sigma_j^m \sigma_k^m , \qquad \langle \sigma_i \sigma_j \sigma_k \rangle = \sum_{\sigma} \sigma_i \sigma_j \sigma_k \, p(\boldsymbol{\sigma}) . \qquad (6.46)$$

If $\mu_{ijk} = \langle \sigma_i \sigma_j \sigma_k \rangle$ for all possible triplets $i, j, k = 1, \cdots, L$, the generated samples will have the same three-point statistics of the training samples. If the same is true for all higher-order moments (four-points, five-points, ...), then the artificial samples are truly statistically indistinguishable from the training samples. Of course, computing higher-order moments is hard in practice. On the one hand the number of possible triplets, quadruplets, etc., grows quickly with $L$, and, on the other hand, the limited availability of test data (*i.e.* small $M$) makes estimation of high-order moments statistically inaccurate. It is generally possible to compute at least the most relevant three-point moments (associated to the largest three-point connected correlations $\mu_{ijk} - \mu_i \mu_{jk} - \mu_j \mu_{ik} - \mu_k \mu_{ij} + 2\mu_i \mu_j \mu_k$) and check if the model reproduces well the empirical moments, see *e.g.* [11] for an application to neural activity data. If some of these moments, or other observables $\mathcal{O}(\boldsymbol{\sigma})$, do not satisfy the matching criterion, then the model can be improved, in the spirit of the maximum entropy principle, by adding the observable to the log-probability of the model with an appropriate Lagrange multiplier, see section 2.3.

One may also directly compare the probabilities of configurations $\boldsymbol{\sigma}$ in the data and in the model. The former can be approximated by the empirical frequencies, and the latter is given by the model distribution $p$. As for moments, estimating frequencies can be quite inaccurate due to sampling limitations. A global criterion for the generative performance of the model, which discounts configurations with low occurrences, is that the average log-likelihoods of test and generated data should have similar values:

$$\frac{1}{M} \sum_{m=1}^{M} \log p(\boldsymbol{\sigma}^m) \simeq \sum_{\sigma} p(\boldsymbol{\sigma}) \log p(\boldsymbol{\sigma}) . \qquad (6.47)$$

Both quantities depend on the hyper-parameters used for learning, such as the intensity $\gamma$ of the regularisation acting on the pairwise couplings, see figure 4.2. It is observed that the value of $\gamma$ such that Eq. (6.47) is satisfied is generally very close to the value of $\gamma$ maximising the log-likelihood of the test set [43]. Hence this global criterion is practically equivalent to the cross-validation procedure of section 4.1.5.

### 6.3.2  Neural-network based formulation of pseudo-likelihood

While pseudo-likelihood maximisation (PLM) was introduced in section 5.3.3 as an inference method for Boltzmann machines, it can be equivalently formulated as the task of predicting a variable $\sigma_i$ (the output) given the other variables $\boldsymbol{\sigma}_{\setminus i}$ (the input). We now rephrase this problem in the language of machine learning[8], in the case of $q = 2$ categories and $\sigma_i = 0, 1$ variables for the sake of simplicity.

For each index $i$ we consider a single-layer neural network with a $L$-dimensional input $\boldsymbol{\sigma}_{\setminus i} = \{\sigma_1, ..., \sigma_{i-1}, 1, \sigma_{i+1}, ..., \sigma_L\}$ connected to a real-valued unit $H_i$ through the weights $\boldsymbol{J}_i = \{J_{i1}, \cdots, J_{i,i-1}, h_i, J_{i,i+1}, ..., J_{i,L}\}$. The value of the unit, given by

---

[8]This is an example of supervised problem, which will be further discussed in Chapter 7.

$H_i = J_i \cdot \sigma_{\backslash i}$ in agreement with Eq. (5.50), is in turn used to determine the output of the neuron, $\sigma_i = 1$ with probability $\Phi(H_i)$ or $\sigma_i = 0$ with probability $1 - \Phi(H_i)$, where

$$\Phi(H_i) = \frac{e^{H_i}}{1 + e^{H_i}} \tag{6.48}$$

is called softmax activation. The parameters $h_i$ and $J_{ij}$, with $j \neq i$ can then be determined by minimising the *cross-entropy loss*, defined as

$$\mathcal{L}_i(J_i) = \frac{1}{M} \sum_{m=1}^{M} \mathcal{C}\left[\sigma_i^m, \Phi\left(J_i \cdot \sigma_{\backslash i}^m\right)\right], \tag{6.49}$$

with

$$\mathcal{C}(\sigma, \Phi) = -\sigma \log \Phi - (1 - \sigma) \log(1 - \Phi). \tag{6.50}$$

The cross-entropy loss is identical to minus the pseudo-loglikelihood defined in Eq. (5.51).

In summary, PLM can be equivalently written in supervised machine learning language as the task of predicting variable $\sigma_i$ given the other variables $\sigma_{\backslash i}$, as solved by an artificial single-layer neural network with weights $J_i$ and softmax activation, being trained via the minimisation of the cross-entropy loss over the $M$ available examples. This new formulation suggests an immediate generalisation of the PLM method based on multilayer neural networks. We replace the conditional probability $\Phi(J_i \cdot \sigma_{\backslash i})$ that the output be $\sigma_i = 1$ in the cross-entropy loss in Eq. (6.49) with

$$\Phi^Q(J_i^Q \cdot \Phi^{Q-1}(J_i^{Q-1} \cdots \Phi^2(J_i^2 \cdot \Phi^1(J_i^1 \cdot \sigma_{\backslash i})))), \tag{6.51}$$

as we did in the case of autoencoders. The activation functions $\Phi^q$ in the intermediate layers can be chosen at will, the only requirement being that the final result should be a number in $[0, 1]$. Their parameters $J_i^q$ are found through loss minimisation. The freedom to play with the number of hidden layer and with the activation and loss functions can significantly improve the expressive power of PLM, of course at the risk of overfitting if the number of trainable parameters becomes too large.

Once learning is carried out configurations $\sigma$ can be sampled from the conditional probabilities as follows. Let us start from a random configuration $\sigma^{(0)}$. Then for each site $i$ we extract a new value of $\sigma_i^{(1)}$ from $p(\sigma_i^{(1)}|\sigma_{\backslash i}^{(0)})$ to obtain a new configuration $\sigma^{(1)}$. We then repeat this step until the statistical properties of $\sigma^{(t)}$ become stationary. This procedure is, however, not guaranteed to converge, and even if it converges, the resulting joint distribution $p(\sigma)$ would be unknown, so the whole sampling is poorly controlled.

### 6.3.3  Autoregressive models

A simple modification of the PLM procedure leads to a class of generative models called autoregressive models. The idea is to first decompose the full joint probability using chain rule for conditional probabilities:

$$p(\sigma) = p(\sigma_1) \times p(\sigma_2|\sigma_1) \times p(\sigma_3|\sigma_2, \sigma_1) \times \ldots \times p(\sigma_L|\sigma_{L-1}, \ldots, \sigma_1). \tag{6.52}$$

This decomposition is exact, and one can then learn a model for each conditional probability $p(\sigma_i|\sigma_{<i})$, where $\sigma_{<i} = (\sigma_1, \ldots, \sigma_{i-1})$. The learning procedure is identical

to PLM, with the only difference that the input is composed of only the variables $\sigma_j$ with $j < i$, instead of all variables $j \neq i$. As in the case of PLM, the simplest choice consists in writing

$$p(\sigma_i = 1|\boldsymbol{\sigma}_{<i}) = \Phi(h_i + \boldsymbol{J}_{<i} \cdot \boldsymbol{\sigma}_{<i}) \,, \qquad (6.53)$$

with the softmax activation $\Phi$ given in Eq. (6.48), and the weight vector defined as $\boldsymbol{J}_{<i} = (J_1, \ldots, J_{i-1})$ *i.e.* a single-layer network. Parameters are then learned by minimising the cross-entropy loss, which is equivalent to maximising the likelihood (and not the pseudo-likelihood) of the data. As for PLM, the expressive power of the model can be increased, if needed, by adding hidden layers and modifying the activation functions.

Note that once the autoregressive model has been trained, one can sample configurations $\boldsymbol{\sigma}$ in linear time: it is enough to first sample $\sigma_1$ from $p(\sigma_1)$, then $\sigma_2$ from $p(\sigma_2|\sigma_1)$, and so on until $\sigma_L$ is sampled from $p(\sigma_L|\sigma_{L-1}, \ldots, \sigma_1)$. This is a great advantage with respect to Boltzmann machines, for which sampling can sometimes be difficult, due to the need of guaranteeing convergence of the Monte Carlo chains. An issue with autoregressive models is, however, that the model depends on the ordering of the variables in the decomposition of the joint probability. The canonical ordering in Eq. (6.52) is not necessarily the most judicious one among the $L!$ possible permutations of the indices of the variables, and empirical criteria exist to select appropriate orderings.

Autoregressive models are versatile and powerful. Important recent applications are the PixelCNN models that achieved impressive performances at image generation [44] and the WaveNet models for raw audio generation [45]. In [46, 47] it has been shown that they can be trained to approximate complex (classical and quantum) statistical mechanics models, while at the same time allowing for easy sampling even when the Monte Carlo dynamics of the original model is very slow.

### 6.3.4 Restricted Boltzmann machines and other architectures

Just like Boltzmann machines, also the restricted Boltzmann machines (RBM) introduced in section 6.2 can be used as generative models. The same kind of statistical tests discussed for Boltzmann machines can be applied to RBMs. The additional advantage of RBMs is that one can not only sample from the full $p(\boldsymbol{\sigma})$, but also from the restricted $p(\boldsymbol{\sigma}|\boldsymbol{h})$ for fixed values of the hidden variables, which is also computationally much simpler, because the variables $\sigma_i$ are decoupled when conditioned to $\boldsymbol{h}$, so no Monte Carlo sampling is required. Hidden variables $\boldsymbol{h}$ may be associated to relevant features of the data, as we have seen in section 6.2.2. Hence, conditioning to $\boldsymbol{h}$ allows one, in principle, to bias the generation process to obtain new data with desired properties.

The field of unsupervised machine learning is in rapid growth, and new architectures are frequently introduced. An example are variational autoencoders (VAE), see [48, 49], which exploit a similar architecture to autoencoders, but with some modifications that turn them into generative models. As in figure 6.1, the encoding part of the network aims at producing internal representations $\boldsymbol{z}$, which are now stochastic. In turn, sampling these internal (or 'latent') variables and passing them through the decoder sub-network produces new data $\boldsymbol{y}'$. This architecture is reminiscent of the so-called Helmholtz machines [50], an extension of RBM in which the encoder and decoder have *a priori* unrelated weights. Training of VAE is done through optimisation of an approximate objective function aiming at maximising the log-likelihood of the decoded data $y' = y$, while making the conditional distribution of latent variables $z$ as close as possible to a simple prior distribution, often chosen to be a product of independent Gaussian variables, with $y$-dependent means and variances. In practice, both the decoder and these means and variances are expressed through multi-layer neural networks. Hence, sampling the latent variables is by construction easy (contrary to the case of RBM) and optimising the objective function, *i.e.* the weights of the encoding/decoding neural networks, can be done by standard backpropagation methods [49]. VAE offer a natural and efficient setup for unsupervised learning, but the arbitrary mapping onto Gaussian distribution may require deep architectures and make interpretation harder. Special efforts are then necessary to produce disentangled representations, *i.e.* relate each component of $z$ to an independent factor of variation of the data.

Another example are generative adversarial networks or GANs [51], which are based on two 'competing' models. Given some training data, a generative model $\mathcal{G}$ tries to capture the data distribution, *i.e.* generate samples that are as similar as possible to the training data, and a discriminative model $\mathcal{D}$ tries to estimate the probability that a sample came from the training data rather than $\mathcal{G}$. The generative model is trained to 'fool' the discriminative model, while the discriminative model is trained to detect the fake data generated by $\mathcal{G}$. The competition between the two adversarial networks ensures that, at the end of a successful training, $\mathcal{G}$ produces data that are hardly distinguishable (even by very performant discriminators $\mathcal{D}$) from the training ones, and is therefore a very good generative model.

### 6.3.5   Application to functional protein design

Among the many possible applications of generative models we illustrate the above methods on one problem of high interest in biology and biogengineering, that is, the design of functional proteins. Proteins are chains of amino acids capable of acquiring complex three-dimensional structures. The subtle pattern of physical-chemical interactions between the 20 types of amino acids make the prediction of the structure and, even more, of the biological function of the protein from the sole knowledge of its sequence extremely difficult.

Over recent years, due to progress in massive sequencing, an increasingly large number of sequences of proteins, coming from widely diverse living organisms have become available. It is therefore possible to collect sequences of proteins having similar structures and functions in organisms having diverged through random mutations for hundreds of millions of years from their last common ancestor, forming a so-called protein family. From a machine learning point of view, sequences are configurations $\sigma$ of $q$-state categorical variables with $q = 21$ values: in addition to the 20 amino-acid types one more category, generally called gap, is introduced to allow for formatting sequences of different lengths, an operation called alignment. We have seen in tutorial 5 how the strong pairwise interactions of a Boltzmann machine learned from those data, *e.g.* through Gaussian inference (used in tutorial 5) or pseudo-likelihood maximisation [52], were informative about the three-dimensional structure of the proteins. This finding is of considerable practical interest as structures are determined from X-ray or NMR experiments from crystallised proteins, a much longer and more costly process than sequencing. Recently, considerable progress in structure prediction was obtained with softwares such as AlphaFold [36] and trRosetta [37], based on deep learning algorithms that are trained both on sequence and structural data. We hereafter consider models that are learned from sequence data only, and we focus on a different problem, namely that of generating artificial functional proteins.

A generative model learned from the sequence data associated to a protein family should be able to produce, via sampling of $p(\boldsymbol{\sigma})$, new artificial protein sequences with the three-dimensional structure and functionality shared by natural sequences in the family. Furthermore, if the features associated to functionalities are well extracted by the model, it should be possible to manipulate them at will, *e.g.* to produce enzymatic proteins with enhanced activity, stability, or even modified substrate specificity.

Learning of BM from protein sequence data provides distributions $p(\boldsymbol{\sigma})$ that reproduce well the frequencies of triplets of amino acids encountered in a protein family [53, 54, 34, 55]. In addition, the models are informative about the outcome of mutagenesis experiments. More precisely, the difference of log-probabilities between a sequence, $\boldsymbol{\sigma}$, and the same sequence with one or multiple mutated amino acids, $\boldsymbol{\sigma}'$, can be used as a predictor of the functional effects of these mutations: positive and negative changes $\log p(\boldsymbol{\sigma}') - \log p(\boldsymbol{\sigma})$ are associated to, respectively, beneficial and detrimental mutations [56, 57].

A more stringent test of the generative power of BM is if entire new sequences $\boldsymbol{\sigma}$, sampled from $p(\boldsymbol{\sigma})$, have the same functionality of the training data in experiments. A recent study of the chorismate mutase protein, an enzyme essential to the synthesis of certain amino acids in bacteria, fungi and plants, seems to indicate this is the case [58]. E.coli bacteria, after removal of their gene coding for chorismate mutase, were unable to grow as expected. However, when provided with synthetic gene sequences generated through Monte Carlo of $p(\boldsymbol{\sigma})$ on plasmids, the bacteria colonies showed normal growth.

Restricted Boltzmann machines were also applied to protein sequence data [59, 60]. The features inferred by the RBM are biologically interpretable, either in terms of structure, or to function, or to phylogenetic identity. Sampling conditionally to some hidden units allows one, for example, to obtain protein sequences with given ligand specificity [59]. Last of all, in [61–63] autoregressive models have been applied to learn generative models of protein families. In particular, in [63] it has been shown that such models provide equivalent performances to BM, but with much lower computational cost for both training and sampling.

## 6.4 Learning from streaming data: principal component analysis revisited

So far we have assumed that data were collected once for all, and then available to the unsupervised learning methods under consideration. In some situations, however, data are produced at very high rate, which makes storage and subsequent processing impossible in practice. How can we adapt unsupervised procedures when data are streamed in real time? In this section, we study this question in the case of principal component analysis (PCA), already introduced in Chapter 3. In its standard offline formulation, PCA extracts relevant directions in the high-dimensional space of data through the diagonalisation of their covariance matrix. We introduce and study below an algorithm to extract principal components from streamed data. The algorithm is said to be online because each data point is 'seen' only once, with no possibility to be revisited or used at later times. While we focus here on PCA for the sake of simplicity, let us emphasise that online learning is a general and important topic. We will further discuss it in the context of supervised learning in Chapter 7.

Furthermore, we will see how the online procedure can be modified to produce sparse outcomes. Principal components are often hard to interpret, as they generally have non-zero projections on all the dimensions of the data space. By demanding that most of these projections vanish, the meaning of the remaining projections may be easier to capture. This sparse PCA procedure is intimately related to autoencoders producing sparse representations, discussed in section 6.1.3.

### 6.4.1 Online learning of the principal component

Consider the simple neural network shown in figure 6.4. A set of $L$ inputs $y_i$, $i = 1, \ldots, L$ are connected to an output $z$ through $L$ weights $J_i$, $i = 1, \ldots, L$. The output is the linear combination of the inputs, $z = \sum_{i=1}^{L} J_i y_i$. The online learning algorithm works in discrete steps, denoted by the integer $t$. Initially (step $t = 0$), the $L$ entries of the weight vector $\boldsymbol{J}^{(0)}$ have very small and random values. At each step $t \geq 1$, an input vector $\boldsymbol{y}^{(t)}$, randomly drawn from the multivariate Gaussian distribution with zero means and covariance matrix $\widehat{C}$, is presented to the input layer, and the corresponding output is computed:

$$z^{(t)} = \sum_{i=1}^{L} J_i^{(t)} y_i^{(t)} = \boldsymbol{J}^{(t)} \cdot \boldsymbol{y}^{(t)} , \qquad (6.54)$$

where a dot denotes the scalar product. The weight vector of the network is then updated according to the learning rule,

$$\boldsymbol{J}^{(t+1)} = \boldsymbol{J}^{(t)} + \eta \, z^{(t)} \left( \boldsymbol{y}^{(t)} - z^{(t)} \boldsymbol{J}^{(t)} \right) , \qquad (6.55)$$

where $\eta$ is a small positive parameter, called learning rate, whose value will be specified later. The learning process is then iterated: a new input $\boldsymbol{y}^{(t+1)}$ is randomly drawn at step $t + 1$ (independently of the previous inputs), the corresponding output $z^{(t+1)}$ is computed and the weight vector is updated to $\boldsymbol{J}^{(t+2)}$, and so on.

Before proceeding, we should keep in mind that:

- Time $t$ can be interpreted as the number of data used by the algorithm. In the following, we will be interested in the limit where $t \to \infty$ at fixed $L$, *i.e.* we have a finite number of parameters to learn using infinite data. This is the ideal setting of asymptotic inference, see Chapter 2.
- However, each data point is used only once to update the weights and then discarded (online setting), while in usual inference one can use all data together (offline setting).

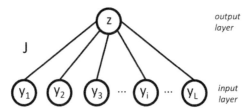

**Fig. 6.4** Illustration of the linear network considered in online PCA.

We are now going to analyse the online learning algorithm defined by Eq. (6.55) and show that the weight vector $J^{(t)}$ converges to the principal component of the correlation matrix $\widehat{C}$. To do so, we use the expression for $z(t)$ in Eq. (6.55) and average[9] the right hand side of Eq. (6.55) over $y^{(t)}$. We obtain

$$J^{(t+1)} = J^{(t)} + \eta \left[ \widehat{C} \cdot J^{(t)} - \left( (J^{(t)})^T \cdot \widehat{C} \cdot J^{(t)} \right) J^{(t)} \right] . \tag{6.56}$$

It is convenient to consider the continuous time limit, in which the learning rate $\eta$ is sent to zero and the number $t$ of steps to infinity with a fixed product $\tau = t \times \eta$ playing the role of continuous (learning) time. In this limit, we have

$$\frac{J^{(t+1)} - J^{(t)}}{\eta} = \frac{J(\tau + \eta) - J(\tau)}{\eta} \to \frac{dJ}{d\tau}(\tau) , \tag{6.57}$$

and the evolution equation for the weight vector becomes

$$\frac{dJ}{d\tau}(\tau) = \widehat{C} \cdot J(\tau) - \left( J(\tau)^T \cdot \widehat{C} \cdot J(\tau) \right) J(\tau) . \tag{6.58}$$

The differential Eq. (6.56) is better expressed in the basis in which the covariance matrix $\widehat{C}$ is diagonal, $\widehat{C}_{ij} = \lambda_i \delta_{ij}$, with eigenvalues $\lambda_1 > \lambda_2 > \ldots > \lambda_L$, with the result

$$\frac{dJ_i}{d\tau}(\tau) = \left( \lambda_i - \sum_j \lambda_j J_j(\tau)^2 \right) J_i(\tau) . \tag{6.59}$$

One can check that, if the weight vector is normalised at some time, say, $\tau = 0$, it remains so at all subsequent times[10].

---

[9]The approximation of replacing the right hand side of Eq. (6.55) by its average is correct in the limit $\eta \to 0$. To show this, one can consider the remainder vector

$$R^{(t)} = z^{(t)} \left( y^{(t)} - z^{(t)} J^{(t)} \right) - \left[ \widehat{C} \cdot J^{(t)} - ((J^{(t)})^T \cdot \widehat{C} \cdot J^{(t)}) J^{(t)} \right] .$$

This remainder is a random variable (as is the input $y^{(t)}$), and has zero mean and finite variance. Hence the sum of the remainder vectors is of the order of $\eta \sqrt{t}$. It will be negligible in the large time $t$, small learning rate $\eta$ limit as long as

$$\eta^2 \times t \to 0 .$$

We will see later on that this is indeed the case.

[10]Consider

$$\frac{1}{2} \frac{d}{d\tau} \sum_i J_i^2 = \sum_i J_i \frac{dJ_i}{d\tau} = \sum_j \lambda_j J_j^2 \left( 1 - \sum_i J_i^2 \right) = 0 .$$

Therefore, if $|J|^2 = 1$, then the derivative of its squared norm is zero, and the normalisation is preserved.

It is straightforward to check that the (normalised) solution of Eq. (6.59) is

$$J_i(\tau) = \frac{J_i(0)\, e^{\lambda_i \tau}}{\sqrt{\sum_j J_j(0)^2\, e^{2\lambda_j \tau}}} \ . \tag{6.60}$$

Extracting the long time behaviour ($\tau \to \infty$) of the formula above, we find[11]

$$J_i(\tau) \xrightarrow[\tau\to\infty]{} \mathrm{sign}(J_1(0))\, \delta_{i1} \ . \tag{6.61}$$

We conclude that $\boldsymbol{J}(\tau)$ converges to the eigenvector associated to the largest eigenvalue, *i.e.* the vector $(\pm 1, 0, 0, 0, 0, \cdots, 0)$, with a random sign depending on the initial condition. Note that the components $J_i(\tau)$ with $i \geq 2$ converge to zero exponentially with rate $\lambda_1 - \lambda_j$. The slowest component is therefore that with $i = 2$, which has a convergence rate

$$\tau_c = \frac{1}{\lambda_1 - \lambda_2} \ , \tag{6.62}$$

that sets the time scale $\tau_c$ needed for the algorithm to converge[12]. Note that the above learning procedure can be extending to extract, say, $K \geq 2$ principal components. The architecture now includes $K$ outputs, with similar learning rules for the $K$ attached weight vectors, and additional "lateral" connections imposing their orthogonalisation [64].

### 6.4.2 Online learning of a sparse principal component

We now consider the same learning rule as in Eq. (6.55) with an extra term in the learning dynamics,

$$\boldsymbol{J}^{(t+1)} = \boldsymbol{J}^{(t)} + \eta \left( z^{(t)} \left( \boldsymbol{y}^{(t)} - z^{(t)}\, \boldsymbol{J}^{(t)} \right) - \gamma \, \mathrm{sign}\left( \boldsymbol{J}^{(t)} \right) \right) , \tag{6.63}$$

where the sign function acts independently on each component, and $\gamma$ is a positive constant. Note that $\mathrm{sign}(0) = 0$ by definition. Rewriting the sign function as

---

[11]The denominator of Eq. (6.60) reads

$$\sqrt{\sum_j J_j(0)^2\, e^{2\lambda_j \tau}} = |J_1(0)| e^{\lambda_1 \tau} \sqrt{1 + \sum_{j=2}^{L} \frac{J_j(0)^2}{J_1(0)^2}\, e^{2(\lambda_j - \lambda_1)\tau}} \simeq |J_1(0)| e^{\lambda_1 \tau} \ ,$$

since eigenvalues are ranked in decreasing order, all the exponential terms in the square root have negative exponent and tend to zero quickly when $\tau \to \infty$. Therefore

$$J_i(\tau) \sim \frac{J_i(0)}{|J_1(0)|}\, e^{(\lambda_i - \lambda_1)\tau} \ ,$$

leading to Eq. (6.61).

[12]In the original discrete time formulation, the learning rate $\eta$ should vary with the number $t$ of steps as

$$\tau = \eta t \gg \frac{1}{\lambda_1 - \lambda_2} \qquad \Leftrightarrow \qquad \eta \gg \frac{1}{t(\lambda_1 - \lambda_2)} \ ,$$

in order to guarantee convergence. This condition is compatible with the condition $\eta^2 t \to 0$ derived in footnote 9.

$$\text{sign}(\boldsymbol{J}) = \nabla_{\boldsymbol{J}} ||\boldsymbol{J}||_1 \, , \tag{6.64}$$

we see that the extra term in the learning rule corresponds to the gradient of a $L_1$-norm regularisation of the weight vector. We expect it will sparsify the principal component, and select only the most relevant $J_i$'s. Following section 6.4.1, we can replace the random right hand side of Eq. (6.63) by its average, and show that in the limit $\eta \to 0$ with fixed product $\tau = t \times \eta$ playing the role of continuous (learning) time, the weight vector obeys the deterministic differential equation

$$\frac{\mathrm{d}\boldsymbol{J}}{\mathrm{d}\tau}(\tau) = \widehat{C} \cdot \boldsymbol{J}(\tau) - \left(\boldsymbol{J}(\tau)^T \cdot \widehat{C} \cdot \boldsymbol{J}(\tau)\right) \boldsymbol{J}(\tau) - \gamma \, \text{sign} \, \boldsymbol{J}(\tau) \, . \tag{6.65}$$

Let us now assume for simplicity that, as in Chapter 3, the correlation matrix $\widehat{C} = \mathcal{I}_L + s \, \boldsymbol{e} \cdot \boldsymbol{e}^T$, with $s > 0$ and $\boldsymbol{e}$ a normalised vector; $\widehat{C}$ has eigenvalues $\lambda_1 = 1 + s$ (associated to $\boldsymbol{e}$), and $\lambda_2 = \lambda_3 = \ldots = \lambda_L = 1$. We introduce two "order parameters":

$$Q(\tau) = \left|\boldsymbol{J}(\tau)\right|^2 \quad \text{and} \quad R(\tau) = \boldsymbol{J}(\tau) \cdot \boldsymbol{e} \, , \tag{6.66}$$

which allow us to rewrite Eq. (6.65) as

$$\frac{\mathrm{d}\boldsymbol{J}}{\mathrm{d}\tau}(\tau) = \left(1 - Q(\tau) - s \, R(\tau)^2\right) \boldsymbol{J}(\tau) + s \, R(\tau) \, \boldsymbol{e} - \gamma \, \text{sign} \, \boldsymbol{J}(\tau) \, . \tag{6.67}$$

*Stability of the zero fixed point.* Recalling that $\text{sign}(0) = 0$, it is clear that $\boldsymbol{J}^* = (0, 0, \ldots, 0)$ is a stationary solution of Eq. (6.67), corresponding to $Q = R = 0$. We now study its stability against a small perturbation, *e.g.* a very small value given to the weight vector at time $\tau = 0$. Keeping only the lowest order terms in Eq. (6.67), we obtain

$$\frac{\mathrm{d}\boldsymbol{J}}{\mathrm{d}\tau} \simeq \boldsymbol{J}(\tau) + s \, R(\tau) \, \boldsymbol{e} - \gamma \, \text{sign} \, \boldsymbol{J}(\tau) \, . \tag{6.68}$$

If $|\boldsymbol{J}| \ll \gamma$, the regularisation term is much bigger than the others, and pushes $\boldsymbol{J}$ to zero, so the stationary point is stable. We thus conclude that the zero weight vector is a locally stable fixed point of the learning dynamics.

*Search for a non-zero fixed point.* Next, we look for non-zero stationary solutions $\boldsymbol{J}^* = (J_1^*, J_2^*, \ldots, J_L^*)$ of Eq. (6.67). We assume $Q + sR^2 - 1 > 0$ and $R > 0$, to be checked self-consistently at the end of the calculation. A careful inspection of this equation lead to the following expression for the fixed point[13]:

---

[13] Two cases must be considered. If $sR|e_i| < \gamma$, then for very small $J_i$ we have

$$\frac{\mathrm{d}J_i}{\mathrm{d}\tau} \approx \begin{cases} sRe_i - \gamma < 0 & \text{for } J_i > 0 \, , \\ sRe_i + \gamma > 0 & \text{for } J_i < 0 \, . \end{cases}$$

Therefore, the only possibility for a stationary solution is $J_i = 0$. Conversely, if $sR|e_i| > \gamma$, then the stationary condition is

$$J_i = \frac{sRe_i - \gamma}{Q + sR^2 - 1} \, , \quad \text{if } s \, Re_i > \gamma \, , \quad \text{and} \quad J_i = \frac{sRe_i + \gamma}{Q + sR^2 - 1} \, , \quad \text{if } s \, Re_i < -\gamma \, .$$

$$J_i^* = \text{sign}(e_i) \times \max\left(0, \frac{s\,R|e_i| - \gamma}{Q + sR^2 - 1}\right). \tag{6.69}$$

Inserting this expression for the weight vector into Eq. (6.66) we obtain two self-consistent equations for $Q$ and $R$:

$$Q = \sum_i (J_i^*)^2 = \sum_i \left[\max\left(0, \frac{s\,R|e_i| - \gamma}{Q + sR^2 - 1}\right)\right]^2,$$

$$R = \sum_i J_i^* \, e_i = \sum_i |e_i| \max\left(0, \frac{s\,R|e_i| - \gamma}{Q + sR^2 - 1}\right). \tag{6.70}$$

### 6.4.3 Discontinuous phase transition in sparse component learning

To illustrate the above findings, let us assume that the top component of the correlation matrix is a sparse vector with a fraction $0 < \rho < 1$ of non-zero components, more precisely,

$$e_i = \begin{cases} \pm\frac{1}{\sqrt{\rho L}} & \text{if } i = 1, \ldots, \rho L \text{ ,} \\ 0 & \text{if } i = \rho L + 1, \ldots, L \text{ .} \end{cases} \tag{6.71}$$

We now ask under what conditions our online learning algorithm is able to recover this component.

Using Eq. (6.69) we see that all the components of $J_i$ vanish if $sR < \gamma\sqrt{\rho L}$. In this case, $R = 0$ according to Eq. (6.66), which of course satisfies the previous condition. Let us now assume that $sR > \gamma\sqrt{\rho L}$. All the components of $J_i$ corresponding to non-zero $e_i$ are non-zero, and we get

$$\boldsymbol{J}^* = \frac{sR - \gamma\sqrt{\rho L}}{Q + sR^2 - 1} \boldsymbol{e} \text{ .} \tag{6.72}$$

Note that $\boldsymbol{J}^*$ is then proportional to $\boldsymbol{e}$. As $\boldsymbol{e}$ is normalized to one, we have

$$R = \boldsymbol{J}^* \cdot \boldsymbol{e} = \frac{sR - \gamma\sqrt{\rho L}}{Q + sR^2 - 1}, \qquad Q = |\boldsymbol{J}^*|^2 = R^2 \text{ .} \tag{6.73}$$

Plugging $Q = R^2$ in the equation for $R$ we obtain

$$R = \frac{s\,R - \Gamma\sqrt{\rho}}{R^2(1 + s) - 1}, \qquad \text{with} \qquad \Gamma = \gamma\sqrt{L} \text{ .} \tag{6.74}$$

We conclude that the scalar product $R$ between the sparse principal component $\boldsymbol{e}$ and the fixed point vector $\boldsymbol{J}^*$ is either zero or solution of the implicit equation Eq. (6.74). As an illustration we show in figure 6.5 the roots $R$ of Eq. (6.74) and the corresponding values of $Q$ in the case $s = 1.7$, $\rho = 0.1$.

As expected, if $\Gamma$ is too large, there does not exist any positive $R$ satisfying Eq. (6.74). A detailed computation shows that $R$ discontinuously jumps to a non-zero value if $\Gamma$ becomes smaller than the critical value

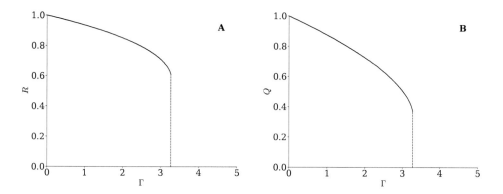

Fig. 6.5 **A.** $R$ versus $\Gamma$ and **B.** $Q$ versus $\Gamma$ for $s = 1.7$ and $\rho = 0.1$.

$$\Gamma_c = \begin{cases} \frac{2(1+s)}{3\sqrt{3}\rho} & \text{if } s \geq 2 \,, \\ \frac{s}{\sqrt{(1+s)\rho}} & \text{if } s \leq 2 \,, \end{cases} \qquad (6.75)$$

and that the value of $R$ at the transition is

$$R_c = \begin{cases} \frac{1}{\sqrt{3}} \approx 0.577 & \text{if } s \geq 2 \,, \\ \frac{1}{\sqrt{1+s}} & \text{if } s \leq 2 \,. \end{cases} \qquad (6.76)$$

We conclude that the modified online learning algorithm can recover a sparse principal component of the form of Eq. (6.71), provided the regularisation $\Gamma = \gamma\sqrt{L} < \Gamma_c$ is not too strong. In the ideal limit of infinite amount of data that we considered here, the overlap $R = \boldsymbol{J} \cdot \boldsymbol{e}$ between the weight vector and the principal component goes to one when $\gamma \to 0$, but of course in the presence of a finite number of data, a finite $\gamma$ is needed to regularise the inference procedure. An application of this sparse PCA algorithm is proposed in tutorial 6.

## 6.5 Tutorial 6: online sparse principal component analysis of neural assemblies

### 6.5.1 Problem

In this tutorial, we are going to use the same dataset as in tutorial 3. As a reminder, the prefrontal cortex activity of a rat learning a task were recorded during $T \sim 23.5$ min. The data were binned over time windows of duration $\Delta t = 100$ ms. The *z-score* variables $y_{bi}$ correspond to $b = 1, \ldots, M = T/\Delta t$ sampled activity states of the $i = 1, \ldots, L = 37$ neurons.

In tutorial 3, we constructed the $L \times L$ Pearson correlation matrix $C = Y^T Y / M$ for the task learning epoch, and computed its principal components. We identified in particular a group of 6 neurons ($i = 1, 9, 20, 21, 26, 27$) that were co-activated during task learning. The data corresponding to the neural recording of the task epoch (*Dati_181014_maze.txt*), and a starting notebook *td6_start.ipynb*, which contains the relevant material extracted from tutorial 3, can be found in the tutorial 6 repository on the book webpage[14].

We would like to repeat the analysis following the online learning procedure of section 6.4. The data $y^{(t)}$ are the *z-score* vectors $(y_{bi}, i = 1, .., L)$ at time $b = t$, where we identify the index $t$ of the online learning time with the index $b$ of the recording time bin.

**Questions:**

1. Write a code that
   - initializes $J^{(0)}$ as a vector whose components are extracted from a Gaussian distribution with mean 0 and variance 1, and which is then normalised,
   - iterates Eq. (6.55) by using once each point in the data set, and
   - returns $u = J^{(t)}/|J^{(t)}|$ at the end the iteration.

   Note that while the fixed point vector is in principle normalised (as shown above), it is safer to enforce normalisation at the end to remove numerical errors.

2. Call $e$ the principal component of the Pearson correlation matrix. Compute the scalar product $R = |e \cdot u|$ between the principal component and its estimate with the online learning procedure of the previous question. Plot $R$ versus $\eta \in [10^{-5}, 10^{-3}]$. Discuss the result. What is the optimal value of $\eta$? Why?
   *Note: the plot will be noisy because of the random initial condition. You might want (optionally) to average results over a few initial conditions.*

3. What happens for $\eta \gtrsim 10^{-2}$? Can you guess why?

4. Call $u^*$ the vector $u$ at the value of $\eta = \eta^*$ that maximises $R$. Plot the square components $(u^*)_i^2$ of the vector $u^*$, identify the largest components and compare with the cell assembly identified in tutorial 3 for the task epoch.

---

[14] https://github.com/StatPhys2DataDrivenModel/DDM_Book_Tutorials

5. Fix $\eta = \eta^*$. Write a code that initialises $\boldsymbol{J}^{(0)} = \boldsymbol{u}^*$, iterates Eq. (6.63) with fixed regularisation parameter $\gamma$, and returns $\boldsymbol{u}' = \boldsymbol{J}^{(t)}/|\boldsymbol{J}^{(t)}|$ together with the norm $Q = |\boldsymbol{J}^{(t)}|^2$ at the end of each iteration.

6. Discuss the dependence of $Q$ on the regularisation parameter $\gamma$ and identify the critical value $\gamma_c$. For $\gamma < \gamma_c$, discuss the behaviour of $\boldsymbol{u}'$.

7. Compare the value of $\gamma_c$ with the theoretical predictions in Eq. (6.75).

8. *Optional* – Instead of using each data point once, repeat the same learning procedure, with a number of passes $\mathcal{N}_{pass} > 1$ over the data set. Set first $\gamma = 0$, and discuss the dependence of the output value for $R$ on $\eta$ and $\mathcal{N}_{pass}$. For what values of $\eta^{**}$ and $\mathcal{N}_{pass}^{**}$ is the principal component recovered with $R > 0.999$? Fix $\eta = \eta^{**}$ and $\mathcal{N}_{pass} = \mathcal{N}_{pass}^{**}$, and use the corresponding normalised component $\boldsymbol{u}^{**}$ as initial condition for the sparse online learning. Repeat questions 4, 5 and 6 above.

### 6.5.2 Solution

1. An implementation of the code is provided in the Jupyter notebook *tutorial6.ipynb*.

2. The overlap $R = |\boldsymbol{u} \cdot \boldsymbol{e}|$ is plotted versus $\eta$ in figure 6.6**A**. For small $\eta$, we are in the continuum limit and the analysis of section 6.4.1 can be applied. The number of data is $t = 14086$ and the total continuous time is $\tau = \eta t$. The condition for convergence is given in footnote 12 above, which with our numerical values $\lambda_1 = 2.15$ and $\lambda_2 = 1.46$, becomes

$$\eta \gg \frac{1}{t(\lambda_1 - \lambda_2)} \approx 10^{-4} \; . \tag{6.77}$$

Indeed, we observe from figure 6.6**A** that $R$ increases from zero (random outcome) to one (perfect overlap) around $\eta \approx 10^{-4}$. However, when $\eta$ becomes too large, the

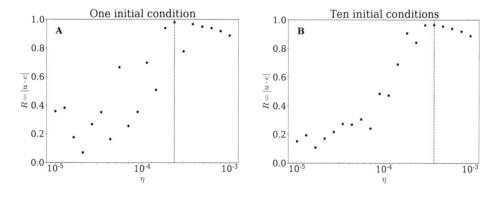

**Fig. 6.6** Overlap $R = |\boldsymbol{u} \cdot \boldsymbol{e}|$ between the principal component and the outcome of online learning. **A**. One single realisation. **B**. Average over 10 realisations.

continuum analysis cannot be applied, and $R$ then decreases again. The plot becomes cleaner when we average over 10 initial random conditions, see figure 6.6**B**. The optimal value is then found to be around $\eta^* \approx 4 \times 10^{-4}$, with some fluctuations depending on the realisation of the initial conditions. From now on we take $\eta^* = 4 \times 10^{-4}$.

3. At even larger $\eta$, the code becomes unstable. The norm $|\boldsymbol{J}|$ increases above 1 and diverges, leading to numerical instability. Recall that this should not happen in the continuum limit, because $|\boldsymbol{J}|$ should tend to one; this effect is then due to the discretisation when $\eta$ is large.

4. If the vector we find were pure noise, we would have $|e_i| \sim 1/\sqrt{L} \sim 0.16$ because the vectors are normalised. In tutorial 3 we used the criterion $|e_i| > 0.2$ to identify the co-activated neurons in offline PCA, and found neurons $\{1, 9, 20, 21, 26, 27\}$. Using the same criterion $|u_i^*| > 0.2$, we find with online PCA (figure 6.7) a slightly different set $\{1, 9, 10, 14, 20, 21, 26\}$ (the values might depend a little bit on the realisation). This suggests that $\{1, 9, 20, 21, 26\}$ is a robust set, while neurons $10, 14, 27$ are sensitive to the noise. Note that in figure 6.7 we plot the squared components of the vector that add up to one.

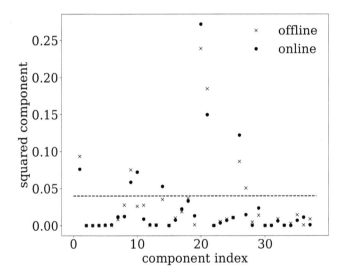

**Fig. 6.7** Squared entries of the first principal component of the Pearson correlation matrix of the activity in the task epoch, obtained from online and offline learning. The largest neuron components identified as co-activated are above the horizontal line.

5. An implementation of the code is provided in the Jupyter notebook *tutorial6.ipynb*.

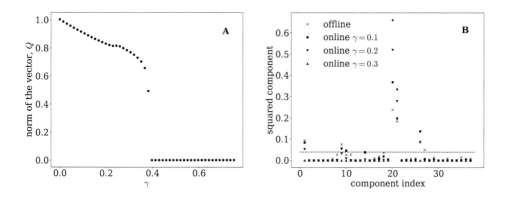

**Fig. 6.8** Results of sparse online PCA. **A**. Norm of the weight vector $Q = |\mathbf{J}|^2$ as a function of regularisation strength $\gamma$. **B**. Examples of weight vectors for several values of $\gamma$.

6. The norm $Q = |\mathbf{J}|^2$ is plotted versus $\gamma$ in figure 6.8**A**. The critical value is identified to be $\gamma_c \approx 0.4$. When $\gamma$ increases, the components of $\mathbf{u}'$ are pushed towards zero. Because $|\mathbf{u}'| = 1$, the non-zero components correspondingly increase. We can assume that the first components to vanish are those associated to noise, while the last one to vanish correspond to neurons that are robustly co-activated. We observe that the components $\{1, 9, 20, 21, 26\}$ are the last to vanish, which confirms our previous identification of this group of neurons.

7. To compare $\gamma_c \approx 0.4$ with the theoretical prediction in Eq. (6.75) we need to estimate $s$ and $\rho$. We have $s = \lambda_1 - 1 \approx 1.15$. We recall that $\rho L$ is the number of non-zero components of $\mathbf{e}$, *i.e.* the number of co-activated neurons; we can estimate this to be $\rho L \approx 5$ according to the above analysis. We then have

$$\gamma_c = \frac{s}{\sqrt{(1+s)\rho L}} \approx 0.35 , \tag{6.78}$$

which provides a decent approximation (given the assumptions involved) of the numerical value.

8. We find that the principal component is recovered with $R > 0.999$ when $\eta = 4 \times 10^{-5}$ and $\mathcal{N}_{pass} = 25$, which guarantees a small enough $\eta$ to be in the continuous limit and a sufficient amount of data to ensure a large enough learning time. Results for $\eta = 4 \times 10^{-5}$ and $\mathcal{N}_{pass} = 25$ are given in figures 6.9 for online learning at $\gamma = 0$ and in figure 6.10 for the sparse case.

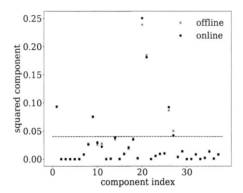

**Fig. 6.9** Same as figure 6.7 but with $\eta = 4 \times 10^{-5}$ and $\mathcal{N}_{pass} = 25$.

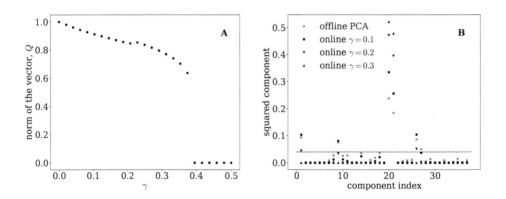

**Fig. 6.10** Same as figure 6.8 but with $\eta = 4 \times 10^{-5}$ and $\mathcal{N}_{pass} = 25$.

# 7

# Supervised learning: classification with neural networks

In Chapter 6 we discussed unsupervised learning as the inference of the underlying statistical structure (and adequate representations) of a set of data. Another fundamental field of application of machine learning is supervised learning, which is the task of learning an input-output relationship. Supervised learning assumes the existence of a function $\sigma(y)$ that associates a label $\sigma$, *e.g.* dog or cat, to each data point $y$, *e.g.* an image. In practice, the machine tries to learn this function from a limited set of input-output examples, *i.e.* pairs $\{y^m, \sigma^m\}$ with $m = 1, \cdots, M$, which have been labelled by a (often, human) supervisor. At the end of learning, one should be able to predict the right class for any vector $y$, not necessarily belonging to the data set: does this new image correspond to a dog or to a cat?

Classification is by no means limited to two classes. The celebrated MNIST data set includes tens of thousands of handwritten digits from 0 to 9; this data base was used by the US postal service to build automatic recognition algorithms capable of reading ZIP codes written on envelopes. We restrict here to two classes only for the sake of simplicity, but most of what follows can be easily extended to any number of classes.

In this chapter, we address two questions:

- How can supervised learning and classification be done in practice?
- How many data does one need to generalise, *i.e.* to avoid overfitting the examples?

We will start by discussing perceptrons, also called support vector machines (SVM), which are neural networks with no hidden layer and have set the state of the art in classification for decades, and we will then turn to more complex neural networks.

## 7.1 The perceptron, a linear classifier

### 7.1.1 Definition of the perceptron network

We consider first the perceptron, a binary classifier, whose architecture is schematised in figure 7.1**A**. The perceptron takes its decision regarding the class $\sigma = \pm$ associated to an input vector $y$ based on a linear combination of the vector components, which is then compared to a threshold $\theta$:

$$\sigma = \text{sign}\left(\sum_{i=1}^{L} J_i\, y_i - \theta\right) .\tag{7.1}$$

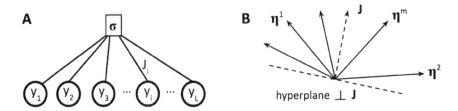

**Fig. 7.1 A.** Architecture of the perceptron and **B**. geometrical interpretation of how it classifies input-output patterns. The perceptron maps an input vector $\boldsymbol{y}$ to a binary output $\sigma$ according to the dot product of $\boldsymbol{y}$ and of the weight vector $\boldsymbol{J}$, see Eqs. (7.1) and (7.2). Given a set of patterns $\boldsymbol{\eta}^m$ defined in Eq. (7.3), learning consists in finding $\boldsymbol{J}$ with a positive scalar product with all these vectors, see Eq. (7.4).

This machine is inspired from the biology of a real neuron, which sums up inputs $y_i$ coming from other neurons $i$ through synaptic connections (represented here by the weights $J_i$), and becomes active or remains silent if the weighted sum exceeds or not a threshold value. It is convenient to add one component to the weight vector, $J_{L+1} = -\theta$, and to each input vector, $y_{L+1} = 1$: the argument of the sign function in Eq. (7.1) then becomes equal to the scalar product of these extended $\boldsymbol{J}$ and $\boldsymbol{y}$ vectors. In the following we will therefore set $\theta = 0$ in Eq. (7.1) without loss of generality.

In a classification problem, we are given $M$ examples of input-output relations $\{\boldsymbol{y}^m, \sigma^m\}$, and look for weights $\boldsymbol{J}$ such that these examples are correctly classified, *i.e.*

$$\sigma^m = \mathrm{sign}(\boldsymbol{J} \cdot \boldsymbol{y}^m) , \qquad \forall m = 1, \ldots, M . \tag{7.2}$$

Let us define the pattern vectors

$$\boldsymbol{\eta}^m = \sigma^m \, \boldsymbol{y}^m . \tag{7.3}$$

The classification conditions expressed in Eq. (7.2) become

$$\boldsymbol{J} \cdot \boldsymbol{\eta}^m > 0 , \qquad \forall m = 1, \ldots, M . \tag{7.4}$$

This requirement has a simple geometrical interpretation sketched in figure 7.1**B**: we look for a vector $\boldsymbol{J}$ such that all the patterns lie on the same side of the hyperplane orthogonal to $\boldsymbol{J}$. As the inequalities (7.4) are invariant under a rescaling of $\boldsymbol{J}$ by a positive factor, we may impose that the weight vector be normalised: $\boldsymbol{J}^2 = 1$.

To make the outcome of the perceptron-based classification more robust, we change the inequality conditions in Eq. (7.4) into

$$\boldsymbol{J} \cdot \boldsymbol{\eta}^m > \Delta , \qquad \forall m = 1, \ldots, M , \tag{7.5}$$

where $\Delta$ is a strictly positive number, hereafter referred to as stability[1]. Under this stronger constraint, variations of the pattern of the type $\boldsymbol{\eta} \to \boldsymbol{\eta} + \boldsymbol{\epsilon}$ will still fulfill Eq. (7.4) provided the norm of $\boldsymbol{\epsilon}$ is small enough. Hence, the outcome of the classification done by the perceptron will be unaffected by the presence of moderate noise in the input vector.

---

[1]This parameter is called margin in the language of support vector machines when the weight vector is normalised.

### 7.1.2 Iterative learning algorithm

Let us assume that a weight vector $\boldsymbol{J}^*$ satisfying the inequalities in Eq. (7.5) exists. We define the largest possible stability,

$$\min_m \boldsymbol{J}^* \cdot \boldsymbol{\eta}^m = \Delta^* > 0 , \qquad (7.6)$$

under the normalisation condition $|\boldsymbol{J}^*| = 1$. We now present a simple interative algorithm guaranteed to output an appropriate normalised weight vector $\boldsymbol{J}$ in a bounded number of steps:

---

**Perceptron learning algorithm**

- Set $\boldsymbol{J}^{(0)} = \boldsymbol{0}$, $t = 0$, and let $c$ be a strictly positive real number.
- Repeat as long as $\min_m \boldsymbol{J}^{(t)} \cdot \boldsymbol{\eta}^m < c$
  - □ Define $m_t = \operatorname*{argmin}_m \boldsymbol{J}^{(t)} \cdot \boldsymbol{\eta}^m$;
  - □ Update $\boldsymbol{J}^{(t+1)} \leftarrow \boldsymbol{J}^{(t)} + \boldsymbol{\eta}^{m_t}$, $t \leftarrow t + 1$.
- Output $\boldsymbol{J} = \boldsymbol{J}^{(t)} / |\boldsymbol{J}^{(t)}|$.

---

Informally speaking, at each step $t$, the algorithm looks for the least stable pattern, *i.e.* having the smaller scalar product with $\boldsymbol{J}^{(t)}$, and pushes the weight vector in its direction at step $t+1$ unless some criterion on the minimal acceptable scalar product is met. After this iteration, the least stable pattern obviously has a larger scalar product with the weight vector. As we see below these greedy moves eventually lead to a perfect classification of all patterns in the dataset.

We will establish the following two statements:

1. The learning algorithm stops after $t_{max} = \dfrac{\eta^2 + 2c}{(\Delta^*)^2}$ steps at most, where $\eta^2 = \max_m |\boldsymbol{\eta}^m|^2$.

2. The stability offered by the output vector $\boldsymbol{J}$ is guaranteed to be a fraction of $\Delta^*$. More precisely, we have $\min_m \boldsymbol{J} \cdot \boldsymbol{\eta}^m \geq \dfrac{c}{\eta^2 + 2c} \Delta^*$.

**Proof of statement one.** Let us define $A^{(t)} = \boldsymbol{J}^* \cdot \boldsymbol{J}^{(t)}$ and $B^{(t)} = |\boldsymbol{J}^{(t)}|^2$. Initially, $A^{(0)} = B^{(0)} = 0$. Then,

$$A^{(t+1)} = A^{(t)} + \boldsymbol{J}^* \cdot \boldsymbol{\eta}^{m_t} \geq A^{(t)} + \Delta^* \geq t\,\Delta^* , \qquad (7.7)$$

where the inequalities holds by virtue of Eq. (7.6) and of the initial condition $A^{(0)} = 0$. Furthermore, as long as the algorithm is running,

$$B^{(t+1)} = B^{(t)} + \left(\boldsymbol{\eta}^{m_t}\right)^2 + 2\,\boldsymbol{J}^{(t)} \cdot \boldsymbol{\eta}^{m_t} < B^{(t)} + \eta^2 + 2c < t\,(\eta^2 + 2c) , \qquad (7.8)$$

where we have exploited the initial condition $B^{(0)} = 0$, and the fact that the halting criterion is not met at step $t$.

The cosine of the angle between $\boldsymbol{J}^{(t)}$ and $\boldsymbol{J}^*$ is therefore given by

$$\frac{\boldsymbol{J}^{(t)} \cdot \boldsymbol{J}^*}{|\boldsymbol{J}^{(t)}|} = \frac{A^{(t)}}{\sqrt{B^{(t)}}} > \frac{\Delta^* \sqrt{t}}{\sqrt{\eta^2 + 2c}} \ . \tag{7.9}$$

As this quantity is bounded from above by one, the algorithm necessarily stops after a number of steps $t_{stop}$ smaller than $t_{max}$, as announced.

**Proof of statement two.** Because the algorithm halts after $t_{stop}$ steps, the stability provided by the output weight vector is

$$\Delta \equiv \min_m \frac{\boldsymbol{J}^{(t_{stop})} \cdot \boldsymbol{\eta}^m}{|\boldsymbol{J}^{(t_{stop})}|} \geq \frac{c}{\sqrt{B^{(t_{stop})}}} \ . \tag{7.10}$$

Using the upper bound over $B^{(t_{stop})}$ in Eq. (7.8) and then the inequality $t_{stop} \leq t_{max}$ we obtain a proof of our second statement:

$$\Delta \geq \frac{c}{\sqrt{\eta^2 + 2c}} \frac{1}{\sqrt{t_{stop}}} \geq \frac{c}{\eta^2 + 2c} \Delta^* \ . \tag{7.11}$$

Hence, as the halting threshold $c$ is sent to infinity, our argument shows that the final stability is at least half $\Delta^*$. A more detailed analysis, which we do not reproduce here, shows that the algorithm actually reaches the optimal stability $\Delta^*$ when $c \to \infty$.

### 7.1.3    Non-linear classification and kernel methods

The version of the perceptron we have studied above allows one to classify data points that are linearly separable; however, many data sets may not be in this situation, as illustrated in figure 7.2. It is possible to extend the perceptron to handle non-linear classification tasks, as we show below.

The key idea is to replace the decision for classification, based on the sign of $\boldsymbol{J} \cdot \boldsymbol{y}$, with

$$\sigma = \text{sign}\big(\tilde{\boldsymbol{J}} \cdot \tilde{\boldsymbol{\phi}}(\boldsymbol{y})\big) \ , \tag{7.12}$$

where $\tilde{\phi}$ is a non-linear function of $\boldsymbol{y}$, of dimension $D$, possibly much larger than $L$. Hence, we keep the linearity of the perceptron, which computes a weighted linear combination of features $\tilde{\phi}_i(\boldsymbol{y})$, $i = 1, \ldots, D$, to take its decision. However, the features are not limited to the $L$ components $y_i$ anymore. For instance, we can include all pairwise (or higher-order) combinations of the components of the input vector,

$$\tilde{\phi}(\boldsymbol{y}) = (y_1, \cdots, y_L, y_1 y_2, y_1 y_3, \cdots, y_{L-1} y_L) \ . \tag{7.13}$$

The weight vector has the same number of components, $D = \frac{1}{2} L(L + 1)$, and reads $\tilde{\boldsymbol{J}} = (J_1, \cdots, J_L, J_{1,2}, J_{1,3}, \cdots, J_{L-1,L})$. According to Eq. (7.12) the input vector $\boldsymbol{y}$ will be assigned to class

$$\sigma = \text{sign}\left( \sum_i J_i \, y_i + \sum_{i<j} J_{i,j} \, y_i \, y_j \right) \ . \tag{7.14}$$

Clearly the representation of the input in Eq. (7.13) is highly redundant, as all the components of $\tilde{\phi}$ can be computed from the $L$ first components only. Going to higher dimension $D \gg L$ may, however, make data linearly classifiable.

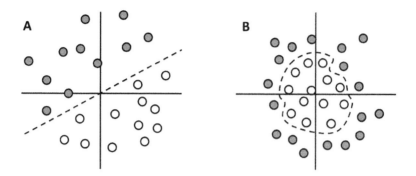

**Fig. 7.2** Classification of data points with linear (**A**) and non-linear (**B**) perceptrons. Each point locates the extremity of the corresponding vector $y^m$, while the colour indicates the class $\sigma^m$. The dashed lines show the decision boundaries between the two classes; for the linear perceptron, this corresponds to the hyperplane orthogonal to $J$, see figure 7.1B. Non-linear classification is based here on a local kernel depending on the distance between its two arguments, such as the Gaussian kernel of Eq. (7.18); the class assigned to a new input is that of the closest data points, see Eq. (7.19).

Linearity of Eq. (7.12) in the $\widetilde{J}_i$'s makes the perceptron algorithm of section 7.1.2 still valid, upon replacement of the patterns $\eta = \sigma\,y$ into $\widetilde{\eta} = \sigma\,\widetilde{\phi}(y)$. Unfortunately, from a computational point of view, the algorithm may become excessively slow when the dimension $D$ is very large, not to specify the memory requirements that grow linearly with $D$.

A clever and computationally efficient implementation of the perceptron classifier and of the learning algorithm for large $D$ can be obtained via the so-called kernel method. As a first observation, we note that we do not explicitly need to compute $\widetilde{J}$ to use the perceptron: only the knowledge of dot products (in the $D$-dimensional space) between $\widetilde{J}$ and the feature vector associated to an input, $\widetilde{\phi}(y)$, is necessary, see Eq. (7.12). We can bypass the storage and computation of these high-dimensional vectors if we know how to swiftly calculate their dot products. A second observation is that the (non-normalised) weight vector after $t$ steps of the perceptron learning algorithm reads

$$\widetilde{J}^{(t)} = \sum_{s=1}^{t} \widetilde{\eta}^{m_s} = \sum_{m=1}^{M} x_m^{(t)}\,\widetilde{\eta}^{m}\ , \qquad (7.15)$$

where $x_m^{(t)} \geq 0$ is the number of time steps $s \leq t$ in which the least stable pattern had index $m_s = m$. Therefore, the determination of the least stable pattern at step $t+1$ as the algorithm is running, or the classification of a new input $y$ once the algorithm has stopped, rely on the computations of dot products of the type

$$\widetilde{J} \cdot \widetilde{\phi}(y) = \sum_{m=1}^{M} x_m\,\widetilde{\eta}^{m} \cdot \widetilde{\phi}(y) = \sum_{m=1}^{M} x_m\,\sigma^m\,\widetilde{\phi}(y^m) \cdot \widetilde{\phi}(y) = \sum_{m=1}^{M} x_m\,\sigma^m\,\kappa(y^m, y)\ ,$$

$$(7.16)$$

where the kernel $\kappa$ is defined through

$$\kappa(\boldsymbol{y}, \boldsymbol{y}') = \tilde{\boldsymbol{\phi}}(\boldsymbol{y}) \cdot \tilde{\boldsymbol{\phi}}(\boldsymbol{y}') \,, \tag{7.17}$$

and maps any two $L$-dimensional input vectors onto a real number. This kernel is the function we really need to be able to evaluate. The $D$-dimensional vectors $\tilde{\boldsymbol{J}}$, $\tilde{\boldsymbol{\phi}}(\boldsymbol{y})$ are merely abstract objects that do not need to be built in practice: the learning algorithm can be entirely described through the updating of the $M$ multiplicities $x_m^{(t)}$, and the classification of a new input $\boldsymbol{y}$ by the perceptron obtained at the end of the training phase depends only on their final values $x_m \equiv x_m^{(t_{stop})}$, see Eq. (7.16). Dealing with $L \times L$-dimensional kernels rather than $D$-dimensional vectors offers a considerable computational advantage if $D$ is very large compared to $L$.

Which kernel function can be chosen? Any kernel $\kappa$ such that Eq. (7.17) holds for some $\tilde{\boldsymbol{\phi}}$ is admissible. The only requirement is therefore that the kernel be a positive matrix[2]. For instance, it can be shown that the following power-law and Gaussian kernels are admissible:

$$k(\boldsymbol{y}, \boldsymbol{y}') = |\boldsymbol{y} \cdot \boldsymbol{y}'|^{\ell} \,, \qquad k(\boldsymbol{y}, \boldsymbol{y}') = \exp\left(-\frac{(\boldsymbol{y} - \boldsymbol{y}')^2}{2\,d^2}\right) \,, \tag{7.18}$$

where $d$ is a free parameter, defining a cutoff distance between $\boldsymbol{y}$ and $\boldsymbol{y}'$. Consider the Gaussian kernel, and suppose we have a set of data points $\{\boldsymbol{y}^m, \sigma^m\}$, see figure 7.2**B**. Once the learning algorithm has output the multiplicities $x_m$, a new point $\boldsymbol{y}$ will be classified according to the rule written in Eq. (7.12):

$$\sigma(\boldsymbol{y}) = \mathrm{sign} \left[ \sum_{\substack{m \text{ s.t.} \\ \sigma^m = +1}} x_m \exp\left(-\frac{(\boldsymbol{y} - \boldsymbol{y}^m)^2}{2\,d^2}\right) - \sum_{\substack{m \text{ s.t.} \\ \sigma^m = -1}} x_m \exp\left(-\frac{(\boldsymbol{y} - \boldsymbol{y}^m)^2}{2\,d^2}\right) \right] \,. \tag{7.19}$$

In words, classifying $\boldsymbol{y}$ with the Gaussian kernel perceptron amounts to compute two weighted sums, running over the examples $m$ in, respectively, the class $\sigma = +1$ and $\sigma = -1$. The term associated to $m$ in each sum is large if the query input $\boldsymbol{y}$ is close to the corresponding input $\boldsymbol{y}^m$ (compared to the cutoff distance $d$) and if the corresponding input was relevant in the course of the training, *i.e.* has large $x_m$. Which sum is larger determines the outcome of the classification, see Eq. (7.19). Briefly speaking, the class of $\boldsymbol{y}$ will be that of the closest, most relevant examples. The locus of the points in the $\boldsymbol{y}$ space where the two sums are equal defines the decision boundary between the two classes, see figure 7.2.

---

[2]This is the case for the linear perceptron, which corresponds to the kernel $k(\boldsymbol{y}, \boldsymbol{y}') = \boldsymbol{y} \cdot \boldsymbol{y}'$. For any function $f(\boldsymbol{y})$, the quadratic form

$$\sum_{\boldsymbol{y}, \boldsymbol{y}'} f(\boldsymbol{y}) \kappa(\boldsymbol{y}, \boldsymbol{y}') f(\boldsymbol{y}') = \sum_{i=1}^{L} \left( \sum_{\boldsymbol{y}} f(\boldsymbol{y}) y_i \right)^2 \geq 0 \,,$$

which shows the positivity of the kernel.

**Fig. 7.3** Dichotomies correspond to linearly separable colourings of $M$ points in a $L$-dimensional space. The black dot locates the origin $\mathbf{O}$ of the space. **A.** Two colourings of $M = 4$ points in dimension $L = 2$. For each colouring, one hyperplane separating the points of each colour is shown. **B.** When a new point $\mathbf{a}$ is added it can be coloured in one or two ways for each colouring of the $M$ points already present.

## 7.2 Case of few data: overfitting

If the dimension $L$ of the perceptron weight vector is large enough compared to the number $M$ of available examples, it is likely that exactly classifying these examples offers no serious guarantees for the correct classification of a new input. To understand how large $M$ should be compared to $L$, we consider the case of purely random input-output pairings, where the class $\sigma^m$ associated to the input vectors $\boldsymbol{y}^m$ is chosen to be $\pm 1$ with equal probabilities $\frac{1}{2}$, independently of $\boldsymbol{y}^m$. Obviously no underlying rule can be reasonably inferred by any classifier in such a situation. The maximal value of $M$ such that the perceptron is able to correctly classify all examples will tell us what should be the minimal size of the dataset necessary to prevent overfitting.

We will follow the geometrical view of classification by a perceptron already presented in figures 7.1**B** and 7.2**A**. Let us put $M$ points $\boldsymbol{y}^m$ in a $L$–dimensional space, and ask in how many ways they can be coloured (with one out of two colours, say, black and white) such that all points of the same colour lie on the same side of a hyperplane, see figure 7.3. Each colouring will be called a dichotomy following the original work by Cover [65].

### 7.2.1 Counting dichotomies

Let $C(M, L)$ be the number of dichotomies over $M$ points in dimension $L$. This number is comprised between 0 and $2^M$. In principle it could depend on the precise positions of the points; actually it does not if the points are generic, *i.e.* any subset of $\leq L$ vectors among the $M$ vectors are linearly independent.

The limit cases are easy to analyse, see figure 7.4:

- $C(1, L) = 2$ for all $L \geq 1$: a single point can always be coloured in black or white.
- $C(M, 1) = 2$ for all $M \geq 1$: in dimension $L = 1$ the unique possible hyperplane separates the points with positive abscissas from the points with negative abscissas. Depending on its orientation, *i.e.* on the sign of the unique component $J_1$, these two subsets of points can be coloured in, respectively, black and white, or vice versa.

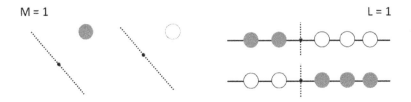

**Fig. 7.4** Enumeration of dichotomies in the limit cases $M = 1$ and $L = 1$.

Let us now establish a recurrence relation on $C(M, L)$. Suppose we have been able to count all the dichotomies on our $M$ points in the $L$-dimensional space. Let us add a new point, say **a**, see figure 7.3**B**. How many dichotomies $C(M + 1, L)$ can we have? This question can be rephrased as: for each dichotomy of the $M$ points, can we move the hyperplane through **a** without changing the colouring of the $M$ points? If we cannot, we get only one dichotomy over the $M + 1$ points given the dichotomy over the $M$ points. If we can, then we get two dichotomies over the $M + 1$ points. This latter situation arises for dichotomies over the $M$ points realized by hyperplanes going through **a** and the origin **0**. Due to this constraint these hyperplanes effectively live in a space of dimension $L - 1$. We expect the number of dichotomies of the projections of the $M$ points onto the subspace orthogonal to (**Oa**) to be equal to $C(M, L - 1)$.

We conclude that

$$C(M + 1, L) = \big[C(M, L) - C(M, L - 1)\big] + 2\, C(M, L - 1)$$
$$= C(M, L) + C(M, L - 1) . \tag{7.20}$$

This recursion relation is identical to Pascal's triangle for binomial coefficients. Together with the boundary cases reported above and in figure 7.4 it allows us to calculate the numbers of dichotomies for all values of $M$ and $L$, with the result[3]

$$C(M, L) = 2 \sum_{i=0}^{L-1} \binom{M - 1}{i} . \tag{7.21}$$

Two consequences of interest are:

- For all $M \le L$ the number of dichotomies is maximal, with $C(M, L) = 2^M$.
- When $M = 2L$ the number of linearly separable dichotomies is exactly half this upperbound: $C(2L, L) = 2^{2L-1}$.

### 7.2.2 The capacity phase transition

We are now ready to tackle the question formulated at the beginning of this section. Suppose we assign uniformly at random each one of $M$ data points in dimension $L$ to the class $-1$ or $+1$, what is the probability that the corresponding dataset can be

---

[3] We use the convention $\binom{k}{l} = 0$ if $l > k$.

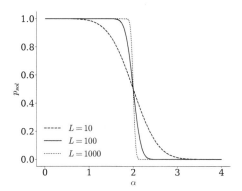

**Fig. 7.5** Probability $p_{sol}(M = \alpha L, L)$ of the existence of a solution for the classification problem with a perceptron for random input-output associations vs $\alpha$ and for sizes $L = 10$, $L = 100$ and $L = 1000$. Notice the abrupt decay from $p_{sol} \simeq 1$ to $p_{sol} \simeq 0$ as $\alpha$ crosses the critical value $\alpha_c = 2$ for large sizes.

linearly classified by a perceptron? This probability, $p_{sol}(M, L)$, is simply related to the number $C(M, L)$ of dichotomies through

$$p_{sol}(M, L) = \frac{C(M, L)}{2^M} , \qquad (7.22)$$

as there are $2^M$ random assignments and only $C(M, L)$ of them are linearly separable. It is plotted in figure 7.5 as a function of the ratio of the number of points over the dimension of the perceptron weight vector,

$$\alpha = \frac{M}{L} , \qquad (7.23)$$

for different values of $L$. As expected $p_{sol}$ is a decreasing function of $\alpha$: it is exactly one for $\alpha \leq 1$, is equal to $\frac{1}{2}$ for $\alpha = 2$, and tends to zero for large ratios $\alpha$.

We also observe that the behaviour of $p_{sol}$ with $\alpha$ becomes sharper and sharper as $L$ increases. Remarkably, in the double infinite size limit, *i.e.* $M, L \to \infty$ with a fixed ratio $\alpha$, the probability that a random class assignment is linearly separable tends to

$$p_{sol}(\alpha) \equiv \lim_{L \to \infty} p_{sol}(M = \alpha L, L) = \begin{cases} 1 \text{ if } \alpha < 2 , \\ 0 \text{ if } \alpha > 2 . \end{cases} \qquad (7.24)$$

The abrupt jump from $p_{sol} = 1$ to $p_{sol} = 0$ signals the onset of a phase transition at the critical ratio $\alpha_c = 2$: for large $L$, a perceptron can almost always classify a set of $(2 - \epsilon)L$ patterns with randomly assigned classes, and can almost never do so for $(2 + \epsilon)L$ patterns. Consequently, for $M < 2L$, we may expect strong overfitting, as even random data can be classified by a perceptron.

In order to better understand this abrupt decay of the probability, we may magnify the vicinity of $\alpha_c = 2$ on a scale decaying with $L$, see figure 7.5. The appropriate width of the neighbourhood of the critical ratio scales as $1/\sqrt{L}$, as shown by the following finite-size scaling formula:

$$\lim_{L\to\infty} p_{sol}\left(L, \alpha = 2 + \frac{a}{\sqrt{L}}\right) \equiv \mathcal{P}(a) = \int_{a/\sqrt{2}}^{\infty} \frac{dt}{\sqrt{2\pi}} \, e^{-t^2/2} \, . \tag{7.25}$$

The function $\mathcal{P}(a)$ smoothly decreases from 1 to 0 as $a$ ranges from $-\infty$ to $+\infty$. The existence of an appropriate scale on which the transition becomes smooth is a characteristic feature of second-order phase transitions in statistical physics[4].

This result obtained by Cover in 1965 [65] was rederived by Gardner in 1988 [66], using techniques from the statistical physics of disordered systems. Gardner's approach is, however, much more general. It can be used to determine how the critical ratio $\alpha_c$ depends on the required stability $\Delta$, or varies with additional constraints on the weights $J_i$, such as $J_i = \pm 1$, for which the direct computation of the number of dichotomies is not possible. It can also be applied to multilayer, non-linear classifiers, see section 7.4. This phase transition is called "capacity" transition for the random perceptron; in different contexts it can also be called a satisfiability transition, because it separates a phase where all constraints (here, the correct classification of a data point) are satisfied from a phase where some constraints are not, see *e.g.* [67].

## 7.3    Case of many data: generalisation

We now analyse the performances of the perceptron when the class $\sigma^m$ of the data points $y^m$ are determined according to a deterministic rule $\tau(y)$. We resort to the so-called teacher-student setting drawn in figure 7.6**A**. A teacher perceptron, with weights $T$, defines the target rule

$$\tau(y) = \text{sign}(T \cdot y) \, . \tag{7.26}$$

We then generate a dataset $\{y^m, \sigma^m = \tau(y^m)\}$, $m = 1, \ldots, M$, where the $y^m$ are random input vectors drawn from an unbiased (isotropic) distribution. These data are used to train a student perceptron with weights $J$ using the learning algorithm of section 7.1.2.

Our aim is to estimate the generalisation error, defined as the probability that the student disagrees with the teacher on a (randomly chosen) test datapoint $y$ (figure 7.6). For given teacher $T$ and student $J$ the generalisation error depends on the angle between the two weight vectors, and reads

$$e_g(J|T) = \text{Proba}\big((J \cdot y) \times (T \cdot y) < 0\big) = \frac{1}{\pi}\text{Arccos}\left(\frac{J \cdot T}{|J|\,|T|}\right) \, , \tag{7.27}$$

---

[4]Consider a $d$-dimensional system exhibiting a second-order phase transition when its volume $V \to \infty$ and some control parameter $\tau$ reaches the critical value $\tau_c = 0$. In this infinite-volume limit the correlation length diverges as $\xi \sim |\tau|^{-\nu}$, where $\nu$ is one of the critical exponents characterizing the transition. For a finite volume $V$ we expect finite-size effects to appear as soon as $\xi$ becomes of the order of the linear size of the system, $\sim V^{1/d}$, hence for $\tau$ of the order of $V^{-1/(d\nu)}$. In the present case, $L$ plays the role of $V$, $\tau = \alpha - \alpha_c$, and $d\nu = 2$ according to Eq. (7.25). The same value is found at the upper critical dimension (at which mean-field theory becomes exact) of the $\phi^4$ theory, since $d_{upper} = 4$ and $\nu_{\text{mean-field}} = \frac{1}{2}$.

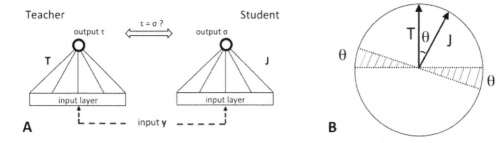

**Fig. 7.6 A.** The generalisation error $e_g$ is defined in the teacher/student framework. The same input $y$ is given to the teacher perceptron, with weight vector $T$, and a student, with weight vector $J$. Their outputs $\tau$ and $\sigma$ are then compared. The fraction of inputs $y$ such that $\tau \neq \sigma$ defines the generalisation error. **B.** Geometrical interpretation of $e_g$, see Eq. (7.27). Inputs $y$ whose projections fall in the hashed area are associated to opposite classes by the student $J$ and the teacher $T$, and contribute to $e_g$. For all the other inputs $y$ both student and teacher assign the same class. We thus have $e_g = (2\theta)/(2\pi)$, where $\theta$ is the angle between $J$ and $T$, see Eq. (7.27). This result is independent on the distribution of $y$ as long as it is isotropic.

as can be seen from figure 7.6**B**. We now want to estimate how this expression varies with $M$ and $L$.

### 7.3.1 Statistical mechanics of student weight vectors

To do so, we consider a statistical mechanics framework in which the $L$–dimensional space of student weight vectors $J$, with $J^2 = 1$, is weighted by the Boltzmann distribution

$$p(J|\{y^m, \sigma^m\})) = \frac{e^{-\beta E(J;\{y^m, \sigma^m\})}}{Z(\{y^m, \sigma^m\})} \ , \tag{7.28}$$

where $\beta$ denotes a fictitious inverse temperature, and

$$E(J; \{y^m, \sigma^m\}) = \sum_{m=1}^{M} \theta(-\sigma^m J \cdot y^m) \ , \tag{7.29}$$

is the "energy". Here, $\theta(u)$ is the Heaviside function, *i.e.* $\theta(u) = 1$ if $u > 0$ and $\theta(u) = 0$ if $u \leq 0$. The energy $E$ measures the training error: it counts the number of misclassified datapoints by the student $J$. The partition function,

$$Z(\{y^m, \sigma^m\}) = \int_{|J|=1} dJ \, e^{-\beta E(J;\{y^m, \sigma^m\})} \ , \tag{7.30}$$

ensures the normalisation of $p$ in Eq. (7.28). When $\beta = 0$, the measure $p$ is uniform over the sphere of unit radius $|J| = 1$. For large $\beta$ it concentrates over the weight vectors that minimise $E$, *i.e.* that correctly classify as many datapoints as possible. We are interested in the generalisation error averaged over the distribution $p$ of students $\boldsymbol{J}$:

$$\langle e_g \rangle = \int_{|\boldsymbol{J}|=1} \mathrm{d}\boldsymbol{J} \, p(\boldsymbol{J}|\{\boldsymbol{y}^m, \sigma^m\}) \, e_g(\boldsymbol{J}|\boldsymbol{T}) \, . \qquad (7.31)$$

This quantity depends on $\boldsymbol{T}$ and the data points $\boldsymbol{y}^m$, and is generally hard to compute. The full calculation of $\langle e_g \rangle$ requires tools and concepts going beyond the scope of this chapter. Yet, deep insights can be gained from the study of the large $M$ regime we consider from now on.

When $M$ is very large, in practice $\gg L$, we expect the training energy $E$ to concentrate around its average over the data points,

$$E(\boldsymbol{J}; \{\boldsymbol{y}^m, \sigma^m\}) \sim M \, e_g(\boldsymbol{J}|\boldsymbol{T}) \, , \qquad (7.32)$$

up to fluctuations of the order of $\sqrt{M}$. The above formula introduces a huge simplication, as it does not explicitly depend on the particular data set under consideration, a property called self-averaging. In this regime, the partition function becomes,

$$Z(M, L) = \int_{|\boldsymbol{J}|=1} \mathrm{d}\boldsymbol{J} \, e^{-\frac{\beta M}{\pi} \mathrm{Arccos}[r(\boldsymbol{J}, \boldsymbol{T})]} \, , \qquad (7.33)$$

where

$$r(\boldsymbol{J}, \boldsymbol{T}) = \boldsymbol{J} \cdot \boldsymbol{T} \, , \qquad (7.34)$$

is the overlap between the student and the teacher weight vectors (both normalised). Because $r$ is a sufficient statistics for $\boldsymbol{J}$, we can rewrite the partition function as

$$Z(M, L) = \int_{-1}^{1} \mathrm{d}r \, e^{-\frac{\beta M}{\pi} \mathrm{Arccos}(r) + S(r)} \, , \quad \text{with} \quad S(r) = \log \int_{|\boldsymbol{J}|=1, \boldsymbol{J} \cdot \boldsymbol{T} = r} \mathrm{d}\boldsymbol{J} \, . \quad (7.35)$$

The first term in the exponent is the effective training energy, which is proportional to $\beta M$. The second term, $S(r)$, is the entropy in the $\boldsymbol{J}$ space associated to a particular value of the overlap $r$ and is expected to scale linearly with the dimension $L$. Hence, depending on the relative scaling of $L$ and $M$, one can be in an entropy-dominated phase for $L \gg \beta M$ where learning is impossible, or in an energy-dominated phase for $L \ll \beta M$, where learning is straightforward. The interesting regime we consider below is when both $L$ and $\beta M$ tend to infinity, with a finite ratio $\tilde{\alpha} = \beta M / L$. As we have assumed that $M \gg L$ to obtain Eq. (7.32) we are implicitly assuming that the temperature is very high (*i.e.* $\beta$ is small): our theory is exact in a regime where we have many data, but each data point influences only weakly the distribution $p$ of student weight vectors $\boldsymbol{J}$.

To compute the integral over $\boldsymbol{J}$ in the entropy $S$ in Eq. (7.35) we decompose $\boldsymbol{J}$ into its components, respectively, parallel and orthogonal to $\boldsymbol{T}$: $\boldsymbol{J} = r\,\boldsymbol{T} + \sqrt{1 - r^2}\,\boldsymbol{J}_\perp$, where $\boldsymbol{J}_\perp$ is a $(L-1)$-dimensional unit vector orthogonal to $\boldsymbol{T}$. Integrating over the $L$-dimensional vector $\boldsymbol{J}$ at fixed $r$ is equivalent to integrating over the $(L-1)$-dimensional sphere $|\boldsymbol{J}_\perp| = 1$. Hence, $S(r)$ is the logarithm of the area of the sphere of radius $\sqrt{1 - r^2}$

in dimension $L - 1$. For large $L$, we have, up to an irrelevant additive constant and to the dominant order in $L$,

$$S(r) = \frac{L}{2} \log(1 - r^2) . \tag{7.36}$$

We end up with

$$\lim_{L \to \infty} \frac{1}{L} \log Z \left( M = \frac{\tilde{\alpha}}{\beta} L, L \right) = \max_{-1 \le r \le 1} \mathcal{F}(r, \tilde{\alpha}) ,$$

$$\text{with } \mathcal{F}(r, \tilde{\alpha}) = \frac{1}{2} \log(1 - r^2) - \frac{\tilde{\alpha}}{\pi} \arccos(r) . \tag{7.37}$$

### 7.3.2 Learning behaviour

The behaviour of $\mathcal{F}(r)$ as a function of $r$ is shown in figure 7.7**A**. There is a unique maximum located in

$$r^*(\tilde{\alpha}) = \frac{\tilde{\alpha}}{\sqrt{\tilde{\alpha}^2 + \pi^2}} , \tag{7.38}$$

moving from $r^* = 0$ when $\tilde{\alpha} = 0$ to $r^* = 1$ when $\tilde{\alpha} \to \infty$. This behaviour confirms that the student weight vector $\boldsymbol{J}$ gets more and more aligned with the teacher weight vector $\boldsymbol{T}$ as the number of data increases. Knowledge of $r^*(\tilde{\alpha})$ gives us the average generalisation error $\langle e_g \rangle = \frac{1}{\pi} \arccos\left(r^*(\tilde{\alpha})\right)$, see Eq. (7.31), with the result shown in figure 7.7**B**.

The decay of the generalisation error at large $\tilde{\alpha}$ is easy to extract analytically. According to Eq. (7.38), we have $r^*(\tilde{\alpha}) \simeq 1 - \pi^2/(2\tilde{\alpha}^2)$ and

$$\langle e_g \rangle \simeq \frac{1}{\tilde{\alpha}} \quad \text{for large } \tilde{\alpha} . \tag{7.39}$$

Interestingly, the average generalisation error also decays as the inverse of the dataset size when training is done at zero temperature ($\beta = \infty$), *i.e.* when the student is asked not to make any misclassification on the training set.

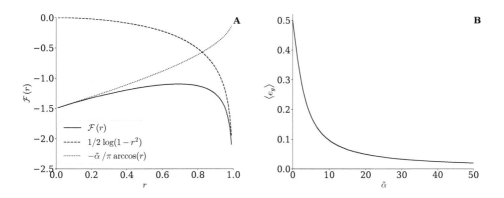

**Fig. 7.7 A.** $\mathcal{F}(r)$ defined in Eq. (7.37) (full line), entropy $S(r)$ (dashed line), and energetic contribution $-\frac{\tilde{\alpha}}{\pi} \arccos(r)$ (dotted line) vs $r$. Note the presence of a unique maximum for $\mathcal{F}$ in $r^*(\tilde{\alpha})$. Parameter value: $\tilde{\alpha} = 3$. **B.** Average generalisation error $\langle e_g \rangle$ vs $\tilde{\alpha}$, decaying as the inverse of the dataset size, see Eq. (7.39).

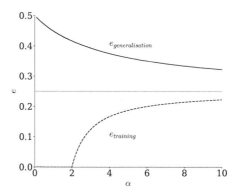

**Fig. 7.8** Sketch of the behaviour of the generalisation and training errors as function of the ratio $\alpha = M/L$ in the case of mismatch between the teacher and student architectures.

The study of the learning of a rule from examples we have just presented can be extended in many directions. Two interesting cases are:

*Mismatch between teacher and student.*    The framework of figure 7.6**A** assumed that both teacher and student were perceptrons. However, in many situations, the two architectures may be different. Of particular relevance is the case of a student less powerful than the teacher: most models we may think of to account for measurements of a real, either natural or artificial, system are prone to be vastly oversimplified.

What can we expect in this mismatch case? Let us assume for definiteness that the student is a perceptron, while the teacher is more complex, such as a multilayered neural network we will study in next section. For small data sets of size $\alpha = M/L \leq 2$ the student perceptron will be able to classify exactly all the examples. The training error will thus vanish. As more examples are fed in to the student, it will not be able to keep up with the teacher's complex rule and the training error will increase until some plateau value is reached, see figure 7.8. This plateau coincides with the average error done by the student on a random input, *i.e.* with the generalisation error. Indeed, the generalisation error is expected to decay from high values ($\frac{1}{2}$) in the strong overfitting regime (small $\alpha$) towards the asymptotic value corresponding to the plateau[5].

*Online learning.*    In the learning procedures we have studied so far, all data are available throughout the learning process. As already discussed in section 6.4, in some situations, data may be streamed for immediate use and cannot be stored. How worse are performances in such an online setting? To make this question more concrete let us assume we have an estimate $\boldsymbol{J}^{(t)}$ of the weight vector of our perceptron after $t$ steps of the algorithm, which now coincides with the number of examples received. As a new example $\left(\boldsymbol{y}^{(t)}, \sigma^{(t)}\right)$ is presented, where the output $\sigma^{(t)} = \text{sign}\left(\boldsymbol{T} \cdot \boldsymbol{y}^{(t)}\right)$

---

[5]The Vapnik-Chernonenkis (VC) dimension, $d_{VC}$, is the maximal value of $M$ such that *all dichotomies* are realisable. In the case of the perceptron, we have shown that $d_{VC} = L$, see discussion following Eq. (7.21). For general architectures of the teacher and student networks of figure 7.6**A** it can be shown that the generalisation and training error made by the student can be made arbitrarily close to one another by increasing the ratio $M/d_{VC}$, where $d_{VC}$ is the VC dimension of the student.

is decided by the perceptron teacher with weight vector $\boldsymbol{T}$, the student weights are updated according to

$$\boldsymbol{J}^{(t+1)} = \boldsymbol{J}^{(t)} + \eta \, f\!\left(\boldsymbol{J}^{(t)} \cdot \boldsymbol{y}^{(t)}\right) \boldsymbol{y}^{(t)} \;, \tag{7.40}$$

where $\eta$ is the learning rate (of the order of $1/\sqrt{L}$), and $f$ is a scalar function. Inspired by the perceptron learning algorithm we first choose

$$f(u) = \begin{cases} 0 & \text{if } \sigma u > 0 \;, \\ \sigma & \text{if } \sigma u < 0 \;. \end{cases} \tag{7.41}$$

Owing to this prescription, $\boldsymbol{J}$ is left unchanged if it correctly classifies the new input, and is pushed in the direction of the misclassified example otherwise.

It can be shown [68] that, with the online learning rule of Eqs. (7.40) and (7.41), the average generalisation error asymptotically decreases as

$$\langle e_g \rangle \propto \frac{1}{\alpha^{1/3}} \quad (\alpha \gg 1) \;, \tag{7.42}$$

a much slower decay than in the offline framework of section 7.3.2, see Eq. (7.39). Remarkably, when the learning rule in Eq. (7.41) is modified into

$$f(u) = \begin{cases} 0 & \text{if } \sigma u > 0 \;, \\ \sigma\,|u| & \text{if } \sigma u < 0 \;, \end{cases} \tag{7.43}$$

the average generalisation error exhibits a much faster decay,

$$\langle e_g \rangle \propto \frac{1}{\alpha} \quad (\alpha \gg 1) \;. \tag{7.44}$$

The reason why this second rule is much more efficient is that it takes into account how badly the misclassified example is, and not only the fact that it is misclassified: $\boldsymbol{J}$ changes only mildly for slightly misclassified examples. It is quite remarkable that this version of online learning may achieve comparable performance to offline learning.

### 7.3.3 When is learning retarded or progressive?

We observe a fundamental difference between the learning curve of the perceptron in figure 7.8 and the one corresponding to principal component analysis (PCA) in figure 3.6C. In both cases, our objective is to infer a $L$-dimensional vector based on a set of $M$ data points. However, extracting the top component requires a minimal number of data $M > M_c = L/\hat{s}^2$, where $1 + \hat{s}$ is the eigenvalue of the covariance matrix associated to this direction. Below this critical number of data, no inference is possible, a phenomenon coined as retarded learning. In the case of the perceptron, however, there is no such critical value, *i.e.* $M_c = 0$. As soon as $M > 0$ generalisation takes place, and the student vector progressively approaches the teacher vector as more and more data are available, and the ratio $\alpha = M/L$ increases. How can we understand this qualitative difference?

To answer this question let us resort to the statistical mechanics framework introduced in section 7.3.1. We expect the density of probability of the overlap $r$ between the target vector (the teacher $T$ in the perceptron case, or the top component $\hat{e}$ in the PCA case) to be given by

$$p(r) \propto e^{S(r)-M\,E(r)} = e^{L\left[\frac{1}{2}\log(1-r^2)-\alpha\,E(r)\right]} , \qquad (7.45)$$

where $S(r)$ represents the entropy in Eq. (7.36) and $E(r)$ plays the role of an energy. In the absence of data, $\alpha = 0$, and $p(r)$ is maximal for $r = 0$ as expected. To understand what happens for positive $\alpha$, we now expand the energy in powers of the overlap $r$:

$$E(r) = E(0) - a\,r - b\,r^2 + O(r^3) , \qquad (7.46)$$

where $a, b$ are real-valued coefficients. In the large $L$ limit, the overlap $r$ becomes concentrated around $r^*(\alpha)$, which maximises

$$\frac{1}{L}\log p(r) = \alpha\,a\,r + \left(\alpha b - \frac{1}{2}\right) r^2 + O(r^3) . \qquad (7.47)$$

Two cases are encountered:

- If $a \neq 0$, we find

$$r^*(\alpha) \simeq 2\,a\,\alpha + O(\alpha^2) . \qquad (7.48)$$

  We expect $a$ to be positive, as the energy should decrease when the similarity with the target direction increases. This is the situation for the perceptron learning problem, where $E(r) = \frac{\beta}{\pi} \text{Arccos } r$ according to Eq. (7.35), and thus $a = \frac{\beta}{\pi}$. Therefore, as soon as $\alpha$ is non zero, so is the overlap $r^*$. This is not surprising as a single example $\sigma^1\,y^1$ is already indicative of the direction of the target $T$. We conclude that, when $a \neq 0$, learning is energy-driven and *progressive*.

- If $a = 0$, we find

$$r^*(\alpha) = 0 \quad \text{if} \quad \alpha < \alpha_c = \frac{1}{2b} , \qquad (7.49)$$

  and that the value of $r^*$ for $\alpha > \alpha_c$ depends of further orders in the expansion of $E(r)$. This situation arises in PCA. Because the ground-truth top component $\hat{e}$ is defined up to a global sign, the sign of the overlap $r$ cannot matter in $E(r)$, hence $a = 0$. In this situation, both the entropy and the energy show the same quadratic behaviour with $r$. For $\alpha < \alpha_c$, the curvature of the entropy being larger (in absolute value) than that of the energy, the maximum of $p(r)$ is located in $r^* = 0$. At the critical ratio $\alpha_c$ the two curvatures compensate each other, and $r^2$ starts to grow for larger ratios. We conclude that, when $a = 0$, learning is entropy-driven and *retarded*.

## 7.4 A glimpse at multi-layered networks

The perceptron is the simplest neural machine: it is composed of an input layer, carrying data configurations, and a binary output. As already seen in the case of unsupervised learning, more complex architectures, include intermediate (or hidden) layers of units can be considered. An example is shown in figure 7.9**A**. Given an input configuration $\boldsymbol{y}$, the values of the $K$ units in the hidden layer are computed according to

$$z_\ell = \text{sign}\left(\sum_{i=1}^{L} J_{\ell,i}\, y_i\right), \qquad \ell = 1,\ldots,K . \tag{7.50}$$

The output unit is then computed through

$$\sigma = \text{sign}\left(\sum_{\ell=1}^{K} W_\ell\, z_\ell\right). \tag{7.51}$$

This multilayered neural network includes the perceptron as a particular case, *e.g.* for $K = 1$ or, equivalently, when all weights but $W_1$ vanish. It can therefore achieve more dichotomies than the perceptron. We expect that, in the double $L, M \to \infty$ limit at fixed ratio $\alpha = M/L$, a randomly chosen dichotomy can be realised with high probability if $\alpha$ is smaller than some critical ratio $\alpha_c(K)$, see figure 7.5. The critical ratio is bounded from above by [69]

$$\alpha_c(K) \le K \log_2 K \quad \text{when} \quad K \gg 1 . \tag{7.52}$$

Consequently, multilayered neural networks promise to offer growing representational capabilities upon increasing the size $K$ of their hidden layer.

The exact value of $\alpha_c(K)$ and, more generally, the generalisation properties of multilayered neural nets are, however, hard to derive, and depend on the detailed architecture of the machines. One well-understood architecture is the so-called committee machine, for which $W_\ell = W > 0$, for $\ell = 1,\ldots,K$: the output $\sigma$ is thus equal to $+1$ if the majority of the hidden units are positive, and to $-1$ if the majority of them are negative. It has been shown that the critical ratio of the committee machine is asymptotically given by

$$\alpha_c(K) = \frac{16}{\pi - 2}\, K \sqrt{\log K} \quad \text{when} \quad K \gg 1 , \tag{7.53}$$

which of course is lower than the upper bound in Eq. (7.52).

The calculation of the generalisation error done for the perceptron in the teacher/student framework of section 7.3 can be extended to the commitee machine. As in the perceptron case, we consider the large number of data, high temperature regime, at fixed ratio $\tilde{\alpha} = \beta M/L$. The typical overlap between the weight vector $\boldsymbol{J}$ connecting one hidden unit to the input layer in the student network and its counterpart $\boldsymbol{T}$ in the teacher machine is again

$$r^*(\tilde{\alpha}) = \underset{r}{\text{argmax}}\, \mathcal{F}(r, \tilde{\alpha}) , \quad \text{with} \quad \mathcal{F}(r, \tilde{\alpha}) = \frac{K}{2} \log(1 - r^2) - \frac{\tilde{\alpha}}{\pi}\, e_g(r) . \tag{7.54}$$

Notice the presence of the factor $K$ multiplying the entropy $S(r)$ defined in Eq. (7.35), as we have to take into account the entropies of the "perceptrons" attached to all the hidden units. Furthermore, the generalisation error $e_g(r)$ is now a more complex function of $r = \boldsymbol{J} \cdot \boldsymbol{T}$ than in the perceptron case. The reason is that $r$ directly

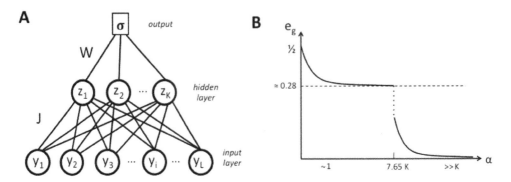

**Fig. 7.9 A.** A multilayer neural network, with one hidden layer composed of $K$ units, whose values are computed as non-linear combinations of the input configuration, with the weights $J$, see Eq. (7.50). In turn the output $\sigma$ is computed from the hidden-layer configuration through the weights $W$, see Eq. (7.51). **B.** Schematic behaviour of the average generalisation error for a committee machine with $K$ hidden units, with the three regimes discussed in the text.

controls the probability $\epsilon(r)$ that the values $z_J$ and $z_T$ of a hidden unit, say, $\ell = 1$, in, respectively, the student and the teacher are different when both networks are given the same input $y$, see Eq. (7.27) and figure 7.6**B**,

$$\epsilon(r) = \mathrm{Proba}(z_J\, z_T = -1) = \frac{1}{\pi} \arccos(r) \ . \tag{7.55}$$

To understand how $e_g$ depends on $r$, we compute the normalised scalar product $\rho$ between the hidden activity vectors $z_J$ of the student and $z_T$ of the teacher, see figure 7.9. As all $K$ hidden units are statistically equivalent, we have, for large $K$,

$$\rho(r) = \frac{z_J \cdot z_T}{|z_J|\,|z_T|} = (+1) \times (1 - \epsilon(r)) + (-1) \times \epsilon(r) = 1 - 2\,\epsilon(r) = \frac{2}{\pi} \arcsin(r) \ . \tag{7.56}$$

Therefore, the generalisation error in Eq. (7.54) depends upon $r$ as follows:

$$e_g(r) = \epsilon\big(\rho(r)\big) = \frac{1}{\pi} \arccos\left(\frac{2}{\pi} \arcsin(r)\right) \ . \tag{7.57}$$

We may now maximise $\mathcal{F}(r)$ over $r$ to obtain the typical overlap $r^*(\tilde{\alpha})$. In the large ratio regime, $\tilde{\alpha} \gg 1$, we have

$$r^*(\tilde{\alpha}) \simeq 1 - \frac{\pi^3}{2}\left(\frac{K}{\tilde{\alpha}}\right)^4 \quad \text{and} \quad e_g \propto \frac{K}{\tilde{\alpha}} \ . \tag{7.58}$$

The formula above shows that the generalisation error strongly decays when $\tilde{\alpha} \gg K$ only. This finding extends to the low temperature regime: $e_g$ decays, for large ratio $\alpha = M/L$, as $K/\alpha$. This result is not surprising. It is reasonable that the number $M$ of data needed for a good generalisation of the teacher rule be large compared to the number $K \times L$ of weights $J_{i\ell}$ in figure 7.9**A**.

A more complete study allows for the derivation of the whole curve of the generalisation error as a function of $\alpha$, with the outcome sketched in figure 7.9**B**. Three regimes can be identified for a committee machine with $K \gg 1$ hidden units:

1. For $\alpha$ of the order of one (with $K \gg 1$), $e_g$ decreases from $\frac{1}{2}$ for $\alpha = 0$ down to a plateau of height $\frac{1}{\pi} \arccos \left( \frac{2}{\pi} \right) \simeq 0.28$ when $\alpha$ is large compared to 1 (but not with respect to $K$). In this initial regime of learning, the weight vector $\boldsymbol{J}$ attached to any hidden unit of the student has equal overlap with all the weight vectors $\boldsymbol{T}$ associated to the teacher hidden units. This regime corresponds to a permutation symmetric phase.

2. As $\alpha$ gets larger than $\simeq 7.65\,K$, a discontinuous change of $e_g$ takes place, corresponding to the breaking of the permutation symmetry between the hidden units. Each hidden unit of the student starts specialising, with a weight vector getting aligned along one weight vector of the teacher. This first order phase transition is sharp in the $M, L \to \infty$ limit only; for finite $L$, fluctuations will randomly destabilise the plateau for smaller values of $\alpha$.

3. As discussed above, for $\alpha \gg K$, the generalisation error decays as the inverse of the dataset size towards zero.

A similar three-fold scenario is also encountered in online learning.

While the discussion in this section was limited to a single multilayer architecture, it is representative of the complexity and richness of behaviour that may be encountered in these machines. Much more remain to be done to understand the generalisation properties of these machines, in particular for structured data.

## 7.5   Tutorial 7: prediction of binding between PDZ proteins and peptides

PDZ protein domains bind small proteins called peptides: each protein has a specificity, *i.e.* it binds some peptides but not others. In [70] the affinity of 74 different PDZ proteins has been tested against $M = 217$ different peptides, by measuring the dissociation constant of the protein-peptide complexes. The aim of this tutorial is to implement a learning algorithm for a perceptron, which for a given PDZ protein, is able to correctly predict the binding of a given peptide sequence of 10 amino acids, *i.e.* associate a binary (binding or non-binding) output to each sequence. We will learn the perceptron on a part of the dataset (training set), and test it on the remaining part (test set).

### 7.5.1   Problem

Consider the peptides in the training set, labelled by $\mu = 1, \ldots, M_{train}$. The sequence of 10 amino acids for the peptide $\mu$ is transformed, through one-hot encoding, in a binary string $s_i^\mu \in \{0, 1\}$, where $i = (n, a)$ is an index associated to the presence of amino acid $a = 1, \ldots, q$ in position $n = 1, \ldots, 10$. Here $q = 19$, because there is no gap symbol and for convenience we choose a gauge symbol that is removed, *e.g.* the last amino acid of the alphabet. The input vector $s^\mu$ has therefore $L = 190$ entries. We can encode the results of the binding assay via a binary output $\sigma^\mu$, which is 1 if the peptide binds and $-1$ if not. We want to find the set of couplings $J_i$ that reproduces the $M_{train}$ input-output associations:

$$\sigma^\mu = \text{sign}\left(\sum_{i=1}^{L} J_i\, s_i^\mu - \theta\right), \qquad \forall \mu = 1, \ldots, M_{train} . \qquad (7.59)$$

For convenience, we introduce a vector of couplings with $L + 1$ components, whose last entry is $J_{L+1} = -\theta$, and input vectors of $L + 1$ components with $s_{L+1}^\mu = 1$. We then multiply the input vector by the output $\sigma^\mu$ to obtain the rescaled input entries $\eta_i^\mu = s_i^\mu \times \sigma^\mu$ for $i = 1, \ldots, L$ and $\eta_{L+1}^\mu = \sigma^\mu$. The input-output equation is then equivalent to the condition:

$$\sum_{i=1}^{L+1} J_i\, \eta_i^\mu = \boldsymbol{J} \cdot \boldsymbol{\eta}^\mu > 0 . \qquad (7.60)$$

If there is a vector of couplings $\boldsymbol{J}$ that solves Eq. (7.60), we are guaranteed to find it by the perceptron learning algorithm discussed in section 7.1.2.

**Data:**
Data are taken from Ref. [70] and can be found in the tutorial 7 repository on the book webpage[6]. The data file *fp_interaction_matrix.xlsx* contains the interaction matrix. It is

---

[6]https://github.com/StatPhys2DataDrivenModel/DDM_Book_Tutorials. The data have been downloaded from the webpage of Ref. [70], https://www.science.org/doi/10.1126/science.1144592.

a Microsoft Excel file, encoding a matrix whose rows are the 217 peptides and whose columns are the 74 PDZs. The entries of the matrix are $-1$ or $0$ if no association between the peptide and the protein is detected in the experiment, and the dissociation constant $k_d$ (in nM), otherwise. The peptide can be considered bound if $k_d < 10^5$ nM. The data file *peptides.free* contains the peptide sequences. Note that the peptides are ordered in different ways in the two data files.

The routine given in the Jupyter notebook *tut7_start.ipynb* can be used to

- read the association matrix and the peptide sequences,
- transform the amino-acid alphabet into numbers,
- construct the $217 \times 191$ matrix $\{s_i^\mu\}$ containing the peptide sequences encoded as above,
- for a given PDZ among the 74 available ones, extract from the data the vector $\sigma^\mu$ that encodes the information about binding.

## Questions:

1. Implement the perceptron learning algorithm described in section 7.1.2, using a large value of $c$. We choose $c = 1000$ in the following.

2. Consider one PDZ (you may start from the 12th protein in the table, *Gm1582 (2/3)*) and use the first $M_{train} = 150$ sequences of the peptides to train the perceptron network. Plot the normalised stability parameter $\boldsymbol{J}^t \cdot \boldsymbol{\eta}^{\mu_t}/|\boldsymbol{J}_t|$ as a function of the number of iterations of the algorithm.

3. Verify that for the test set, composed by the $M_{test} = 217 - M_{train} = 67$ remaining sequences, the network reproduces the input-output association. The performances of binary classifications are usually measured by the following quantities. Define Np=number of positives, Npp=number of predicted positives, Ntp=number of true positives, Nn=number of negatives, Npn=number of predicted negatives, Ntn=number of true negatives. Calculate: *i)* True positive rate or sensitivity: Ntp/Np, *ii)* Precision or positive predictive value: Ntp/Npp, *iii)* True negative rate or specificity: Ntn/Nn, *iv)* Negative predictive value: Ntn/Npn, *v)* Accuracy (Ntp+Ntn)/$M_{test}$. Discuss the obtained results.

4. Change the training set (number of sequences and data set): are results stable with respect to such variations? For example, make a table of results for the 1st, 12th, 18th and 51th protein with the first 120 peptides in the training set, or the first 150 peptides in the training set. You can also randomly select the peptides in the training and test set.

5. *Optional:* use a linear regression with the lasso algorithm to fit the training set. Compare the quality of the predictions on the test set with the perceptron result.

6. *Optional:* use the Keras package to train a perceptron, with a sigmoid activation function and a binary cross-entropy loss (in a $0, 1$ representation of the binding labels). Compare the results with the perceptron learning algorithm.

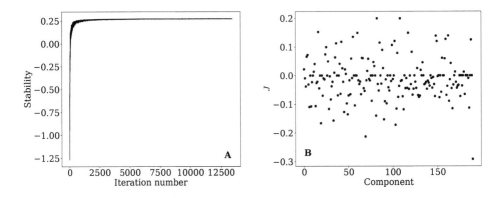

**Fig. 7.10 A**. Stability as a function of the algorithm step for the 12th protein (*Gm1582* (*2/3*)) using as training set the first 150 peptides. **B**. The resulting components of the learned vector **J**.

### 7.5.2   Solution

1. An implementation of the code is provided in the Jupyter notebook *tutorial7.ipynb*.

2. The stability of the network as function of the algorithm step for the 12th protein (*Gm1582* (*2/3*)) using as training set the first 150 peptides is shown in figure 7.10**A**, and the final vector **J** is shown in figure 7.10**B**.

3. In the available data, the non-binding (negative) result is much more frequent than the binding (positive) result. Hence, it is generically rare to find positives in the test data. Results for some selected PDZ are given in table 7.1. The TNR and NPV are quite large, because the predictions are more often a negative than a positive, while the TPR and PPV can be very low, *e.g.* when no positive are predicted and no positive are in the test set. See *e.g.* the first protein (labelled 0) that has only two binding peptides, which are both in the training set and correctly reproduced. Protein number 50 is an exception, because many more positive are predicted than the real number in the test set. The true negative rate is therefore relatively small.

| Protein nb | $M_{train}$ | TPR | PPV | TNR | NPV | Accuracy |
|---|---|---|---|---|---|---|
| 11 | 150 | 0.67 | 0.67 | 0.98 | 0.98 | 0.97 |
| 11 | 120 | 0.16 | 0.2 | 0.95 | 094 | 0.9 |
| 50 | 120 | 0.5 | 0.09 | 0.85 | 0.98 | 0.88 |
| 0 | 150 | 0 | 0 | 1 | 1 | 1 |
| 17 | 150 | 0.5 | 0.16 | 0.8 | 0.96 | 0.82 |

**Table 7.1** Performance of the perceptron inferred from data on the binding (positive)/non-binding (negative) of some selected PDZ proteins with peptides. Here proteins are numbered from 0 to 73. The first $M_{train}$ peptides are used as training set, the remaining ones as test set.

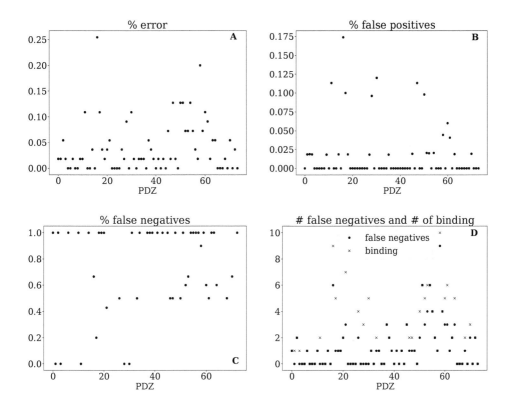

**Fig. 7.11** Performances of the perceptron algorithm on the 74 PDZ proteins (indexed from 0 to 73 in the horizontal axis). The 217 peptides are split in 162 used as training and 55 used as test. Shown are: **A**. the total error on the test set (fraction of incorrectly predicted bindings), **B**. the fraction of false positives and **C**. of false negatives, and **D**. the absolute number of false negatives (unpredicted bindings) compared to the total number of binding peptides.

4. Results are not very stable by changing the training and test sets. Some results for all proteins are given in figure 7.11.

5. Lasso regression generically underperforms the perceptron result. As an example we can consider again the 12th protein (*Gm1582 (2/3)*) using as training set the first 150 peptides. Performing a lasso regression with regularisation parameter $\gamma = 0.001$ we obtain a perfect fit (zero training error) but a test error of 13% (*i.e.* 13% of the test set is misclassified). Increasing regularisation to $\gamma = 0.01$ we obtain a 8% training error and 4.4% test error (the parameters are shown in figure 7.12). By comparison, the perceptron achieves a zero training error and 3.0% test error.

6. The Keras package can train neural networks using gradient descent on a loss function. We used here a perceptron model with sigmoid activation function and a binary cross-entropy loss (gradient descent requires smooth enough activation functions). For

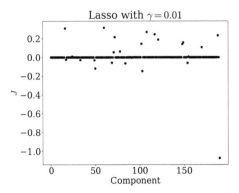

**Fig. 7.12** Components of the vector $J$ obtained from a lasso regression with $\gamma = 0.01$.

the 12th protein *(Gm1582 (2/3))*, using as training set the first 150 peptides, the results are comparable to those obtained via the perceptron algorithm (figure 7.13).

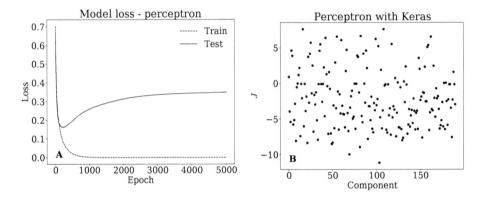

**Fig. 7.13** Results obtained via the Keras package for a perceptron with sigmoid activation and binary cross-entropy loss, on the 12th protein *(Gm1582 (2/3))* using as training set the first 150 peptides. **A.** Loss function computed on the training and test set during training via the Keras package. **B.** Components of the vector $J$ at the end of training.

# 8

# Time series: from Markov models to hidden Markov models

So far, we have considered inference problems in which time played no role[1] . In many applications, however, data are time series produced by a dynamical process. How can we infer the underlying rules defining this process? We will address this question in two frameworks: a simple one, in which measurements give directly access to the dynamical sequence of the states visited by the system, and a richer one, in which we only have indirect knowledge about the states.

## 8.1 Markov processes and inference

### 8.1.1 What is a Markov process?

Let us assume that the system under study evolves according to a discrete time dynamics, and is in state $y_t \in \mathcal{Y}$ at time $t = 0, 1, 2, \dots$. The size of $\mathcal{Y}$, that is, the number of accessible states is $D$. We consider a stochastic dynamical process for the evolution of the system, in which the conditional probability to be in a state $y_{t+1}$ at time $t+1$ only depends on the state $y_t$ at time $t$:

$$p(y_{t+1}|y_t, y_{t-1}, \dots, y_1) = p(y_{t+1}|y_t) \ . \tag{8.1}$$

In other words, this probability is independent of the states visited at times $t' < t$. Such a process is called memoryless or Markovian, and is entirely defined by the transition matrix

$$\Omega(y \to y') = p(y'|y) \ , \tag{8.2}$$

whose elements are the conditional probabilities[2] that the system visits state $y'$ at time $t+1$ given it was in state $y$ at time $t$. The entries of the transition matrix satisfy two important properties:

1. $\Omega(y \to y') \geq 0$ for all $y, y' \in \mathcal{Y}$, and
2. $\sum_{y' \in \mathcal{Y}} \Omega(y \to y') = 1$ for all $y \in \mathcal{Y}$;

the second normalisation condition expresses that the system must visit one of the states at the next time, irrespectively of the state it is in. The non-zero elements

---

[1]This was also the case when learning from streamed data in sections 6.4 and 7.3.2. Any permutation in the order in which data were presented would have led to the same machine.

[2]We are also assuming that the conditional probabilities are independent of time, *i.e.* the transition matrix has no explicit dependence on time.

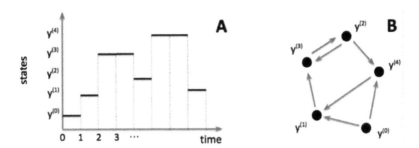

**Fig. 8.1** Example of time evolution of the state of a system (**A**) generated by a Markov process (**B**). As time goes on the system may hop from one state to another, with probabilities defined by the transition matrix $\Omega$, see Eq. (8.2).

of $\Omega$ define the set of possible transitions, which can be observed along trajectories (figure 8.1)

The above description can be easily generalised to continuous-time evolution processes, by denoting $\Omega(y \to y') \geq 0$ the rate, *i.e.* the probability per unit time, of going from $y$ to $y'$. This rate is positive for all $y' \neq y$, and the normalisation condition imposes $\Omega(y \to y) = -\sum_{y' \in \mathcal{Y}} \Omega(y \to y')$, which can be interpreted as minus the decay rate of the probability to stay in state $y$. Notice also that all finite-time memory processes can be turned into a Markov process by increasing the dimension of the states. If $K$ measures the time horizon, *i.e.*

$$p(y_{t+1}|y_t, y_{t-1}, \ldots y_1) = p(y_{t+1}|y_t, y_{t-1}, \ldots, y_{t-K-1}) , \tag{8.3}$$

then it is easy to convince oneself that $\hat{y}_t = (y_t, y_{t-1} \ldots y_{t-K+1})$ undergoes a Markovian dynamics.

### 8.1.2 Inference of the transition matrix

Suppose now that we observe a trajectory $(y_0, y_1, \ldots y_M)$, such as in figure 8.1**A**. How can we infer the transition probabilities $\Omega(y \to y')$ in figure 8.1**B**?

Let us start by writing the probability of this trajectory (given the initial state),

$$p(y_1, \ldots, y_M|y_0) = \prod_{t=1}^{M} \Omega(y_{t-1} \to y_t) = \prod_{y,y'} \Omega(y \to y')^{\mathcal{N}(y \to y')} , \tag{8.4}$$

where we have reordered the terms in the product over the different times, and

$$\mathcal{N}(y \to y') = \sum_{t=1}^{M} \delta_{y,y_{t-1}} \, \delta_{y',y_t} \tag{8.5}$$

is the number of transitions from $y$ to $y'$ along the trajectory. These transition numbers form a sufficient statistics of the data.

The maximum likelihood estimator (MLE) for the $\Omega$ matrix is obtained by maximising the log-probability of the trajectory in Eq. (8.4), under the constraint that the sum of elements along each column must be equal to one:

$$\Omega^{MLE} = \underset{\Omega}{\operatorname{argmax}} \left\{ \sum_{y,y'} \mathcal{N}(y \to y') \, \log \Omega(y \to y') - \sum_y \lambda(y) \left( \sum_{y'} \Omega(y \to y') - 1 \right) \right\} ,$$
(8.6)

where the $\lambda$'s are Lagrange multipliers enforcing the constraints. Upon differentiation with respect to $\Omega$ we obtain

$$\frac{\mathcal{N}(y \to y')}{\Omega^{MLE}(y \to y')} = \lambda(y) ,$$
(8.7)

for all $y, y'$. The solution of this equation satisfying the normalisation constraint reads

$$\Omega^{MLE}(y \to y') = \frac{\mathcal{N}(y \to y')}{\sum_z \mathcal{N}(y \to z)} .$$
(8.8)

Hence, the MLE estimator for $\Omega$ is simply the frequency of transitions observed in the data. A simple regularisation scheme consists in introducing pseudo-counts for transitions that are not observed, which is useful for inference made from short trajectories.

### 8.1.3 Markov processes at equilibrium

For well-chosen transition matrices, the Markov process is guaranteed for long enough times to reach a stationary distribution, $p^{eq}(y)$. One sufficient condition[3] is detailed balance, which expresses that the oriented fluxes of probability between any pair of states $y, y'$ compensate:

$$p^{eq}(y) \, \Omega(y \to y') = p^{eq}(y') \, \Omega(y' \to y) .$$
(8.9)

For a trajectory of duration $M$ much larger than the equilibration time[4] we have

$$\sum_z \mathcal{N}(y \to z) \approx M \, p^{eq}(y) , \qquad \mathcal{N}(y \to y') \approx M \, p^{eq}(y) \, \Omega(y \to y') ,$$
(8.10)

up to relative errors of the order of $1/\sqrt{M}$. Combining Eqs. (8.8) and (8.10) shows the statistical consistency of maximum likelihood inference: $\Omega^{MLE} \to \Omega$ when $M \to \infty$.

---

[3] Another requirement is that the chain be ergodic: there must exist, for any pair of states $y_{start}$ and $y_{end}$, a path in $Y$ starting in $y_{start}$ and ending in $y_{end}$ such that all transitions along the path are admissible (have non-zero $\Omega$).

[4] The largest eigenvalue of $\Omega$ is equal to one, and the corresponding eigenmode is $p^{eq}$. The second largest eigenvalue $\lambda_2 < 1$ controls the equilibration time, through $\lambda_2 = e^{-1/t^{eq}}$.

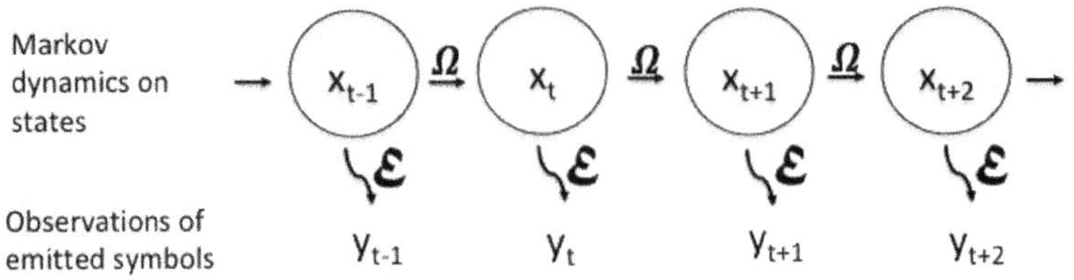

**Fig. 8.2** Schematic representation of a hidden Markov model: the dynamics over the hidden states represented by the $x_t$ units is Markovian and defined by the transition matrix $\Omega$, but only the symbols $y_t$ emitted by these states according to the conditional probability $\mathcal{E}$ are observables.

We may also compute the log-probability of the trajectory per data point,

$$-\frac{1}{M} \log p(y_1, ..., y_M | y_0) = -\frac{1}{M} \sum_{y,y'} \mathcal{N}(y \to y') \log \Omega(y \to y')$$

$$= \sum_y p^{eq}(y) \left( -\sum_{y'} \Omega(y \to y') \log \Omega(y \to y') \right), (8.11)$$

where the last line is obtained when $M \to \infty$. We conclude that the log-probability of the trajectory per observation is asymptotically equal to the conditional entropy of $p(y'|y)$ given $p^{eq}(y)$, also called entropy rate of the Markov process.

## 8.2 Hidden Markov models

We have seen in section 8.1 how to infer the transition matrix of a Markov process from the observation of temporal trajectories of its states. Unfortunately, in many contexts, the states are not directly accessible. Measurement noise can alter the observations, or the relation between what is measured and the underlying dynamical states defining the system is not straightforward. An adequate framework to deal with such a situation is provided by hidden Markov models (HMM). HMM have a huge range of applications, in signal processing, speech recognition, automatic translation, alignments of proteins and RNA sequences, gene analysis, and so on; see *e.g.* [32, 33].

### 8.2.1 Definitions and problems

Figure 8.2 shows the operation of a HMM. As time $t$ increases, the state of the system, now denoted by $x_t$, evolves according to a Markov process defined by the transition matrix $\Omega$. At each time step, a symbol $y_t$ is emitted at random according to the conditional probability $\mathcal{E}(y_t | x_t)$. The data are composed by the time sequence of symbols $(y_1, y_2, ..., y_M)$. Contrary to section 8.1 states are not accessible from the data: they are hidden, or latent, and manifest themselves indirectly through the emitted symbols only.

Formally, a HMM is defined by

- the set of possible states $x^{(i)}$, $i = 1, \ldots, D$, and the $D \times D$ transition matrix from one state to another, $\Omega(x^{(i)} \to x^{(i')})$;

- the set of possible symbols $y^{(j)}$, $j = 1, \ldots, Q$, and the $Q \times D$ emission matrix $\mathcal{E}(y^{(j)}|x^{(i)})$ defining the conditional probability of a symbol given a state.

We consider only the case of finite sets of states and symbols, but the extension to infinite and/or continuous sets is straightforward[5]. Notice that neither $\Omega$ nor $\mathcal{E}$ vary with time.

There are three problems of increasing difficulties we are interested in:

- **Problem 1.** Given a transition matrix $\Omega$, an emission matrix $\mathcal{E}$, and an initial state $x_0$, how can we compute the probability $p(\boldsymbol{y})$ of a sequence of observed symbols $\boldsymbol{y} = (y_1, y_2 \ldots y_M)$?
- **Problem 2.** Given $\Omega$, $\mathcal{E}$, what is the conditional probability $p(\boldsymbol{x}|\boldsymbol{y})$ of a sequence of hidden states, $\boldsymbol{x} = (x_1, x_2 \ldots x_M)$, given a sequence of observed symbols $\boldsymbol{y}$? In particular what is the trajectory of states $\boldsymbol{x}^{MLE}(\boldsymbol{y})$ with highest probability?
- **Problem 3.** Given a sequence of symbols $\boldsymbol{y}$, how can we infer $\Omega$, $\mathcal{E}$?

We explain below how these problems can be solved.

### 8.2.2  Solution to problem 1: the transfer matrix formalism

To compute the probability of the sequence $\boldsymbol{y}$ we have to look at all possible sequences of states $\boldsymbol{x}$: each $\boldsymbol{x}$ has its own probability to be generated by the Markov process and to generate the sequence of symbols. We write

$$
p(\boldsymbol{y}) = \sum_{x_1, x_2, \ldots x_M} p(x_1, \ldots, x_M | x_0) \prod_{t=1}^{M} \mathcal{E}(y_t | x_t)
$$

$$
= \sum_{x_1, x_2, \ldots x_M} \prod_{t=1}^{M} \Omega(x_{t-1} \to x_t) \prod_{t=1}^{M} \mathcal{E}(y_t | x_t) . \tag{8.12}
$$

At first sight this formula looks useless as it requires a summation over a huge number, $D^M$, of sequences of states. However, with a clever trick introduced in statistical physics in 1940 by Wannier under the name of transfer matrix, and known in computer science as dynamic programming [71], the summation can be done with a computational effort scaling only linearly in $M$. Let us introduce the $D \times D$ matrices $\mathcal{M}_t$, whose elements are defined by

$$
\mathcal{M}_t(x', x) \equiv \Omega(x \to x') \, \mathcal{E}(y_t | x') . \tag{8.13}
$$

Notice that $\mathcal{M}_t$ explicitly depends on the time $t$ through the symbol $y_t$.

---

[5] A very important subset of HMM are defined by both $\Omega$ and $\mathcal{E}$ being Gaussian, and the states and symbols being continuous vectors (with a priori distinct dimensions). The dynamical evolution rules are

$$
\boldsymbol{x}_{t+1} = A \, \boldsymbol{x}_t + \boldsymbol{w}_t , \quad \boldsymbol{y}_t = B \, \boldsymbol{x}_t + \boldsymbol{v}_t ,
$$

where $A$ and $B$ are, respectively, square and rectangular matrices, and $\boldsymbol{w}_t$ and $\boldsymbol{v}_t$ are Gaussian white noises. The methods developed in the 60's to deal with this specific case of HMM are known under the name of Kalman filters.

The probability of the sequence of symbols can be rewritten as follows:

$$p(\boldsymbol{y}) = \sum_{x_1,\dots,x_M} \mathcal{M}_M(x_M, x_{M-1}) \times \dots \times \mathcal{M}_t(x_t, x_{t-1}) \dots \times \mathcal{M}_2(x_2, x_1) \times \mathcal{M}_1(x_1, x_0)$$

$$= \sum_{x_M} \left[ \mathcal{M}_M \cdot \dots \cdot \mathcal{M}_t \cdot \dots \mathcal{M}_2 \cdot \mathcal{M}_1 \right](x_M, x_0) \,. \tag{8.14}$$

Let us emphasise that the terms in the first line of Eq. (8.14) are products ($\times$) of the elements of the matrices $\mathcal{M}_t$, while the terms in the second line are elements of the matrix product ($\cdot$) of the same $\mathcal{M}_t$'s. To be more precise, in the second line of Eq. (8.14) we exploit the definition of the matrix product,

$$\sum_{x_{t-1}} \mathcal{M}_t(x_t, x_{t-1}) \times \mathcal{M}_{t-1}(x_{t-1}, x_{t-2}) = \left[ \mathcal{M}_t \cdot \mathcal{M}_{t-1} \right](x_t, x_{t-2}) \,, \tag{8.15}$$

and iterate this relation for all times $t \le M$.

We have therefore reduced the computation of $p(\boldsymbol{y})$ to that of the product of $M$ different matrices, which can be carried out in time $M \times D^3$. In practice, we only need $M \times D^2$ computations as we just have to compute the iterates of the vector representing the initial state, $\boldsymbol{v}_0 = (0, \dots, 0, 1, 0, \dots, 0)$, where the position of the 1 corresponds to $x_0$. This is a huge gain with respect to the naive summation over the $D^M$ sequences in Eq. (8.12).

### 8.2.3   Solution to problem 2: the Viterbi algoritm

We now turn to the estimation of the best sequence of states given an observed sequence of symbols. Using Bayes' rule the probability of any sequence state $\boldsymbol{x}$ given $\boldsymbol{y}$ can be expressed as

$$p(\boldsymbol{x}|\boldsymbol{y}) = \frac{p(\boldsymbol{y}|\boldsymbol{x}) \times p(\boldsymbol{x})}{p(\boldsymbol{y})} = \frac{\mathcal{M}_M(x_M, x_{M-1}) \times \dots \times \mathcal{M}_t(x_t, x_{t-1}) \times \dots \times \mathcal{M}_1(x_1, x_0)}{\sum_{x_M} \left[ \mathcal{M}_M \cdot \dots \cdot \mathcal{M}_1 \right](x_M, x_0)} \,. \tag{8.16}$$

This probability can be easily computed in a time linear in $M$ for any sequence[6]. However, there is again an exponential-in-$M$ number of sequences of states, and determining which one has the highest probability by exhaustive enumeration is simply impossible.

Fortunately, the Markovian nature of the dynamical process allows us to drastically simplify the search for the best sequence of states. The key observation is that a state, say, $x_t$, appears in two matrix elements at the numerator in Eq. (8.16) only, $\mathcal{M}_{t+1}(x_{t+1}, x_t)$ and $\mathcal{M}_t(x_t, x_{t-1})$. If we fix the values of $x_{t-1}$ and $x_{t+1}$ the optimisation over $x_t$ is easy to perform. This observation can be turned into an efficient algorithm, with a running time linear in $M$ and $D$, and invented by Viterbi [72].

---

[6]It is also straightforward to compute marginal distributions over the states using the transfer matrix formalism of Eq. (8.14). As an illustration the joint distribution of states at time $t$ and $t'$ (with $t' > t$) reads

$$p(x_t, x_{t'}) = \frac{\sum_{x_M} \left[ \mathcal{M}_M \cdot \dots \cdot \mathcal{M}_{t'+1} \right](x_M, x_{t'}) \times \left[ \mathcal{M}_{t'} \cdot \dots \cdot \mathcal{M}_{t+1} \right](x_{t'}, x_t) \times \left[ \mathcal{M}_t \cdot \dots \cdot \mathcal{M}_1 \right](x_t, x_0)}{\sum_{x_M} \left[ \mathcal{M}_M \cdot \dots \cdot \mathcal{M}_1 \right](x_M, x_0)} \,.$$

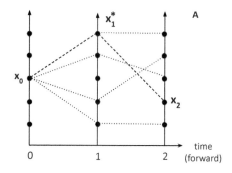 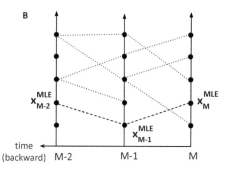

**Fig. 8.3** Schematic descriptions of the forward (**A**) and backward (**B**) procedures in the Viterbi algorithm. Each dot represents a state $x$ (here, $D = 5$). **A**. Times $t = 0, 1, 2$. The dashed line shows the best path connecting $x_0$ to $x_2$, which goes through $x_1^*$ at time $t = 1$. The dotted lines show the best paths corresponding to other choices of the state at time $t = 2$. **B**. Times $t = M - 2, M - 1, M$. The dashed line shows the reconstruction of the optimal (MLE) path, from $x_M^{MLE}$ at time $M$, and then backwards in time. This optimal path is obtained by recursive applications of the relations $x_{t-1}^*(x_t)$ shown by the dotted lines.

*Forward procedure.* Suppose we want to find the best subsequences $(x_0, x_1, x_2)$, joining the known initial state $x_0$ to each possible $x_2$. There are $D$ such optimal paths, which are shown in figure 8.3**A**. As $x_0$ is known and $x_2$ is fixed, the optimal state at time $t = 1$ is given by

$$x_1^*(x_2) = \operatorname*{argmax}_{x_1} \ell_2(x_2, x_1) \quad \text{with} \quad \ell_2(x_2, x_1) = \log \mathcal{M}_2(x_2, x_1) + \log \mathcal{M}_1(x_1, x_0) .$$
(8.17)

This optimal state depends on $x_2$ (and on $x_0$ that is fixed once for all). As we choose another ending state $x_2$ for our length-3 path the optimal $x_1$ varies, see figure 8.3**A**.

Let us now look for the best length-4 subsequences $(x_0, x_1, x_2, x_3)$, joining the known initial state $x_0$ and each possible $x_3$. Given the final state $x_3$, the optimal sub-sequence will go through some optimal states $x_2$ and $x_1$. As we have already determined the optimal state $x_1^*$ as a function of $x_2$ in Eq. (8.17), the optimal $x_2$ can be determined through

$$x_2^*(x_3) = \operatorname*{argmax}_{x_2} \ell_3(x_3, x_2) \quad \text{with} \quad \ell_3(x_3, x_2) = \log \mathcal{M}_3(x_3, x_2) + \ell_2\big(x_2, x_1^*(x_2)\big) .$$
(8.18)

Iterating this procedure forward in time allows us to compute the best state as a function of the next one, $x_t^*(x_{t+1})$, as well as the corresponding log product of matrix elements, $\ell_{t+1}$.

*Backward procedure.* Once we have reached $t = M - 1$ and computed $x_{M-1}^*(x_M)$ we are left with determing the optimal state $x_M$. This is simply done as follows

$$x_M^{MLE} = \operatorname*{argmax}_{x_M} \ell_M\big(x_M, x_{M-1}^*(x_M)\big) .$$
(8.19)

We may now go backward in time to determine the best path, as illustrated in figure 8.3**B**. The optimal state at time $M - 1$ is

$$x_{M-1}^{MLE} = x_{M-1}^* (x_M^{MLE}) , \qquad (8.20)$$

and so on using $x_{t-1}^*(x_t)$ until we obtain the full sequence of states. The log-probability of this optimal path is

$$\log p(x^{MLE}|y) = \ell_M (x_M^{MLE}, x_{M-1}^{MLE}) - \log p(y) , \qquad (8.21)$$

where $p(y)$ is given in Eq. (8.14).

We have assumed so far that the initial state $x_0$ was known. If this is not the case, it can be inferred with the Viterbi algorithm, with the difference that the forward pass starts in $t = 1$ and computes $x_0^*(x_1)$, and the backward pass ends in $t = 1$ and provides $x_0^{MLE}$.

### 8.2.4   Solution to problem 3: the expectation-maximisation procedure

The last and hardest problem we want to solve is the inference of the parameters $\theta = (\Omega, \mathcal{E})$ that define the HMM from data, that is, a sequence of symbols $y$. We would like to do so through the maximisation of the likelihood $p(y|\theta)$ in Eq. (8.14).

Because the likelihood can be efficiently calculated, so can its gradient with respect to $\theta$. Therefore, a simple, local learning algorithm is gradient ascent of the likelihood. This algorithm is guaranteed (if the learning rate is small enough) to converge towards a local maximum of the likelihood. Resuming the procedure from another initial value for $\theta$ allows one to span the space of parameters, and find better and better (higher and higher) local maxima.

Unfortunately, all currently known algorithms for HMM learning are not fundamentally better than gradient ascent, in that they do not guarantee convergence towards the global maximum of $p(y|\theta)$. We are going to present one of these algorithms, developed by Baum and Welch in the 60s [73]. The algorithm starts from some guess for the value of the parameters, $\theta_0$, and then updates it through discrete steps, $\theta_0 \rightarrow \theta_1 \rightarrow \theta_2 \rightarrow \dots$. Though the steps may not be local, *i.e.* $\theta_{i+1}$ is not necessarily close to $\theta_i$, the likelihood increases after each step, until a local maximum is reached. This algorithm illustrates a quite general method of inference, called expectation-maximisation (EM).

*Objective function for the EM procedure.*   Given the current estimate $\theta_i$ of the model parameters we define

$$G_i(\theta) = \sum_x p(x|y, \theta_i) \log p(x, y|\theta) , \qquad (8.22)$$

where $p(x, y|\theta)$ is the conditional joint distribution of $x$ and $y$ given $\theta$. $G_i$ is a good objective function, meaning that

$$\theta_{i+1} = \underset{\theta}{\text{argmax}} \, G_i(\theta) \qquad (8.23)$$

is a better estimate of the model parameters than $\theta_i$:

$$p(\boldsymbol{y}|\boldsymbol{\theta}_{i+1}) \geq p(\boldsymbol{y}|\boldsymbol{\theta}_i) \;. \tag{8.24}$$

To establish Eq. (8.24), we use the chain rule for conditional probabilities[7], and rewrite

$$\log p(\boldsymbol{y}|\boldsymbol{\theta}) = G_i(\boldsymbol{\theta}) - \sum_{\boldsymbol{x}} p(\boldsymbol{x}|\boldsymbol{y}, \boldsymbol{\theta}_i) \log p(\boldsymbol{x}|\boldsymbol{y}, \boldsymbol{\theta}) \;. \tag{8.25}$$

This equality is true for all $\boldsymbol{\theta}$, including $\boldsymbol{\theta}_i$. Substracting Eq. (8.25) for $\boldsymbol{\theta} = \boldsymbol{\theta}_i$ from the same equation for generic $\boldsymbol{\theta}$, we obtain

$$\log p(\boldsymbol{y}|\boldsymbol{\theta}) - \log p(\boldsymbol{y}|\boldsymbol{\theta}_i) = G_i(\boldsymbol{\theta}) - G_i(\boldsymbol{\theta}_i) + \sum_{\boldsymbol{x}} p(\boldsymbol{x}|\boldsymbol{y}, \boldsymbol{\theta}_i) \log \left[ \frac{p(\boldsymbol{x}|\boldsymbol{y}, \boldsymbol{\theta}_i)}{p(\boldsymbol{x}|\boldsymbol{y}, \boldsymbol{\theta})} \right] \;. \tag{8.26}$$

The last term is the Kullback-Leibler divergence $D_{KL}(p(\boldsymbol{x}|\boldsymbol{y}, \boldsymbol{\theta}_i)||p(\boldsymbol{x}|\boldsymbol{y}, \boldsymbol{\theta}))$, which is positive. Therefore,

$$\log p(\boldsymbol{y}|\boldsymbol{\theta}) - \log p(\boldsymbol{y}|\boldsymbol{\theta}_i) \geq G_i(\boldsymbol{\theta}) - G_i(\boldsymbol{\theta}_i) \;. \tag{8.27}$$

With the choice $\boldsymbol{\theta} = \boldsymbol{\theta}_{i+1}$ defined in Eq. (8.23) the right hand side of Eq. (8.27) is positive, yielding Eq. (8.24). Hence, the log-likelihood increases after each step of the algorithm. We emphasise that this procedure, based on the maximisation of $G_i(\boldsymbol{\theta})$ in Eq. (8.22), is generic and can be applied to any inference problem.

*EM procedure for HMM.* Let us now see how the maximisation of $G_i(\boldsymbol{\theta})$ is done in the specific case of HMM. We have, after $i$ steps of the learning algorithm, $\boldsymbol{\theta}_i = (\Omega_i, \mathcal{E}_i)$ and

$$p(\boldsymbol{x}, \boldsymbol{y}|\boldsymbol{\theta}_i) = \prod_{t=1}^{M} \Omega_i(x_{t-1} \to x_t) \prod_{t=1}^{M} \mathcal{E}_i(y_t|x_t) \;. \tag{8.28}$$

According to Eq. (8.22) we want to optimise

$$G_i(\Omega_i, \mathcal{E}_i, \mu_i, \eta_i) = \sum_{\boldsymbol{x}} p(\boldsymbol{x}|\boldsymbol{y}, \boldsymbol{\theta}_i) \sum_{t=1}^{M} \left( \log \Omega_i(x_t \to x_{t+1}) + \log \mathcal{E}_i(y_t|x_t) \right) \tag{8.29}$$

$$- \sum_{x} \lambda_i(x) \left( \sum_{x'} \Omega(x \to x') - 1 \right) - \sum_{x} \eta_i(x) \left( \sum_{y} \mathcal{E}(y|x) - 1 \right) \;,$$

where we have introduced the $D$-dimensional Lagrange multipliers $\lambda_i(x)$ and $\eta_i(x)$ to enforce the normalisation conditions over, respectively, the transition and the emission matrices.

We may now differentiate $G_i$ in Eq. (8.29) with respect to the transition matrix elements, and impose that the derivatives should vanish at the maximum. Similarly to

---

[7]In the case of three variables, this chain can be expressed as follows:

$$p(y_1, y_2|y_3) = \frac{p(y_1, y_2, y_3)}{p(y_3)} = \frac{p(y_1, y_2, y_3)}{p(y_2, y_3)} \times \frac{p(y_2, y_3)}{p(y_3)} = p(y_1|y_2, y_3) \times p(y_2|y_3) \;,$$

which simply follows from the definition in Eq. (1.7).

the calculation done in section 8.1.2, we find, after solving for the Lagrange parameters $\lambda_i(x)$,

$$\Omega_{i+1}(x \to x') = \frac{\langle \mathcal{N}(x \to x') \rangle_{\boldsymbol{\theta}_i}}{\sum_z \langle \mathcal{N}(x \to z) \rangle_{\boldsymbol{\theta}_i}} \,, \tag{8.30}$$

where $\mathcal{N}(x \to x')$ is the number of transitions observed in trajectories of states generated by the distribution $p(\boldsymbol{x}|\boldsymbol{y}, \boldsymbol{\theta}_i)$ defined by the current parameters $\boldsymbol{\theta}_i$, and $\langle \cdot \rangle_{\boldsymbol{\theta}_i}$ denotes the average over this distribution. The maximisation of $G_i$ in Eq. (8.29) with respect to the emission matrix elements can be done in the same way, with the result

$$\mathcal{E}_{i+1}(y|x) = \frac{\langle \mathcal{S}(x,y) \rangle_{\boldsymbol{\theta}_i}}{\sum_{y'} \langle \mathcal{S}(x,y') \rangle_{\boldsymbol{\theta}_i}} \,, \tag{8.31}$$

where $\mathcal{S}(x,y)$ is the number of times the system in is state $x$ and the symbol $y$ is observed (in simulated data).

The expression for $\Omega_i$ in Eq. (8.30) is thus formally similar to Eq. (8.8), with the major difference that the averages $\langle \cdot \rangle_{\boldsymbol{\theta}_i}$ are computed from plausible data inaccessible to the measurements, generated according to the current model with parameters $\boldsymbol{\theta}_i$, and not over accessible data as was the case in section 8.1. The calculations of the averages $\langle \cdot \rangle_{\boldsymbol{\theta}_i}$ constitutes the *expectation step*. The frequencies of transitions then provide, through Eq. (8.30), the new estimate of the transition matrix (*maximisation step*). The same EM procedure applies to the emission matrix through Eq. (8.31).

## 8.3 Tutorial 8: CG content variations in viral genomes

The $\lambda$-phage is a virus that infects bacteria. The phage genome, which contains 48502 bases, has the particularity to display different regions with different CG base pair contents. Because the base pairing energy between AT base pairs is smaller than the one for CG base pairs, AT-rich regions separate (or denaturate) at lower temperature than CG-rich regions. It is thought that the difference in CG content in different regions of the $\lambda$-phage genome may be due to the need to quickly denaturate the DNA that is inserted in the infected bacteria.

### 8.3.1   Problem

**Preliminary calculations:**

- **Identification of AT-rich or CG-rich regions in the $\lambda$-phage genome.** In order to identify AT-rich and CG-rich regions one can build a hidden Markov model, in which the emission probability of one of the four symbols at position $i$ along the sequence depends on a hidden state, representing the property of $i$ being in an AT-rich region or in a CG-rich region. The probability of emission of a letter ($y_i \in$ {A,C,G,T}) in position $i$ is therefore written as $\mathcal{E}(y_i|x_i)$ where the hidden variable is $x_i = 0$ for AT-rich regions and $x_i = 1$ for CG-rich regions. The emission probabilities $\mathcal{E}(y_i|x_i)$ and transition probabilities $\Omega(x_i \to x_{i+1})$ are taken from Ref. [32, figure 4.2], where they have been derived as the best HMM that fits the $\lambda$-phage sequence data. The emission probabilities of a base $y_i \in$ {A,T,C,G} extracted from the $\lambda$-phage sequence in both AT-rich and CG-rich regions are given in table 8.1, and the probability of transition between the two hidden states $x \in \{0, 1\}$ along the sequence are:

$$\Omega(0 \to 1) = \Omega(1 \to 0) = 2 \cdot 10^{-4} , \tag{8.32}$$

  while the probabilities to stay in the same hidden state are

$$\Omega(0 \to 0) = 1 - \Omega(0 \to 1) , \qquad \Omega(1 \to 1) = 1 - \Omega(1 \to 0) . \tag{8.33}$$

- **Most probable sequence of hidden states along the $\lambda$-phage genome.** In order to find the most likely hidden sequence we make use of the Viterbi algorithm described in section 8.2.3. In this setting, a sequence of emitted symbols $y_i$ is known, and one wants to reconstruct the most likely sequence of hidden symbols $x_i^*$, that maximises the HMM probability. Suppose that we know the most likely sequence $\{x_0^*, \ldots, x_{i-1}^*\}$ up to site $i - 1$. The log-probability for site $i$ can be written as

$$\ell_{i+1}(x_i, x_{i+1}) = v(x_0^*, \ldots, x_{i-1}^*, x_i) + \log \mathcal{E}(y_i|x_i) + \log \Omega(x_i \to x_{i+1}) , \tag{8.34}$$

  where, assuming that the initial state $x_0$ has probability $p_{in}(x_0) = 1/2$, we have $v(x_0) = \log(1/2)$ and for $i > 0$ we also have

|     | AT-rich | CG-rich | Average |
|-----|---------|---------|---------|
| A   | 0.2700  | 0.2462  | 0.2581  |
| C   | 0.2084  | 0.2476  | 0.228   |
| G   | 0.1981  | 0.2985  | 0.2483  |
| T   | 0.3235  | 0.2077  | 0.2656  |

**Table 8.1** Emission probabilities $\mathcal{E}(y|x)$ in AT-rich ($x = 0$) and CG-rich ($x = 1$) regions. Their average is also given for reference. Data taken from Ref. [32, figure 4.2].

$$v(x_0^*, \ldots, x_{i-1}^*, x_i) = \log(1/2) + \sum_{n=0}^{i-1} \log \mathcal{E}(y_n | x_n^*)$$

$$+ \sum_{n=0}^{i-2} \log \Omega(x_n^* \to x_{n+1}^*) + \log \Omega(x_{i-1}^* \to x_i) .$$
(8.35)

From this expression, we can find the sequences of hidden states $x_i$ conditioned to the following one $x_{i+1}$. We first find the most probable hidden state at position $i$ along the sequence for each of the two possible hidden states in position $i + 1$, $x_{i+1} = 0$ or 1, by maximising Eq. (8.34) with respect to $x_i$:

$$x_i^*(x_{i+1}) = \text{argmax}_{x_i} \, \ell_{i+1}(x_i, x_{i+1})$$
(8.36)

At the last position $L - 1$ along the sequence we find $x_{L-1}^*$ by omitting the last term in Eq. (8.34), and we then backpropagate to obtain the most likely sequence $x_{L-2}^*(x_{L-1}^*), x_{L-3}^*(x_{L-2}^*), \ldots, x_0^*(x_1^*)$.

## Questions:

1. Read the data[8] file *lambdaphage.txt* and convert the ACGT alphabet from letters to numbers $y \in \{0, 1, 2, 3\}$: use the routine which is given in the Jupyter notebook *tut8_start.ipynb* to have the nucleotide alphabet. Both files are available in the tutorial 8 repository on the book website[9].

2. The probabilities of observing each base in a AT-rich ($x = 0$) or CG-rich ($x = 1$) sequence are given in table 8.1. Decide, by a Bayesian inference method, which of the two is the best model to fit the first and second half of the $\lambda$-phage sequence.

3. Consider a sliding window of fixed size $k$, *i.e.* the sites $(i, i + 1, \cdots, i + k - 1)$, and compute the percentage of G and C bases on the sequence, averaged over the $k$ sites along the sequence. Plot this quantity as a function of the central point $i + k/2$, for some values of $k$ (e.g. $k = 10, 100, 200, 1000$) and compare with Ref. [32, figure 4.3].

4. For fixed $k$, decide, by the same Bayesian decision method of question 2, if the subsequence of the sites $(i, i + 1, \cdots, i + k - 1)$ is CG-rich or AT-rich. Plot the result as a function of $i + k/2$ and add it to the plot of point 3.

[8] The file has been downloaded from https://www.ncbi.nlm.nih.gov/nuccore/215104.
[9] https://github.com/StatPhys2DataDrivenModel/DDM_Book_Tutorials

5. Determine the sequence $x_i^*$ of the best hidden variables along the sequence and the transition points between AT-rich and CG-rich regions using the Viterbi algorithm on the HMM described above. Add $x_i^*$ to the plot of point 3, and compare with Ref. [32, figure 4.3].

### 8.3.2   Solution

2. Let us consider, in a Bayesian inference setting, a given sequence $\boldsymbol{y} = (y_1, \ldots, y_k)$ of observed symbols. We want to perform a *hypothesis testing*, to decide whether the sequence comes from a AT-rich region ($x = 0$) or a GC-rich region ($x = 1$). We start by writing the probability of the data given $x$. Assuming independent emissions, we have

$$p(\boldsymbol{y}|x) = \prod_{i=1}^{k} \mathcal{E}(y_i|x) \ . \tag{8.37}$$

Because we have no prior information on $x$, we can assume a uniform prior $p(x) = 1/2$. Hence, under Bayes' rule, the probability of $x$ given the observed sequence is $p(x|\boldsymbol{y}) \propto p(\boldsymbol{y}|x)$. We decide that hypothesis $x = 1$ is correct if $p(x = 1|\boldsymbol{y}) > p(x = 0|\boldsymbol{y})$, so the proportionality factor is irrelevant. We can define

$$R(\boldsymbol{y}) = \log \frac{p(x = 1|\boldsymbol{y})}{p(x = 0|\boldsymbol{y})} = \sum_{i=1}^{k} \log \frac{\mathcal{E}(y_i|x = 1)}{\mathcal{E}(y_i|x = 0)} = \sum_{i=1}^{k} \beta(y_i) \ , \tag{8.38}$$

with

$$\beta(y) = \log \frac{\mathcal{E}(y|x = 1)}{\mathcal{E}(y|x = 0)} \ . \tag{8.39}$$

Note that it is convenient to take the logarithm to avoid multiplying many numbers, which results in exponential behaviour and possible numerical problems. A positive $R(\boldsymbol{y})$ indicates a GC-rich sequence ($x = 1$), while a negative $R(\boldsymbol{y})$ indicates a AT-rich sequence ($x = 0$). According to this analysis, the best model for the first part of the sequence (first 25000 base pairs) is GC-rich ($x = 1$), while for the second part of the sequence (last $\sim$ 25000 base pairs) the best model is the AT-rich model ($x = 0$).

3. The percentage of G and C base pairs, averaged on a window of $k = 200$ or $k = 1000$ sites, is plotted in figure 8.4.

4. The result of the Bayesian decision on a window of $k = 200$ or $k = 1000$ sites is plotted in figure 8.4.

5. The sequence of best hidden variables along the $\lambda$-phage DNA sequence, with the HMM parameters given above, is plotted in figure 8.4. It displays three CG-rich regions and four AT-rich regions.

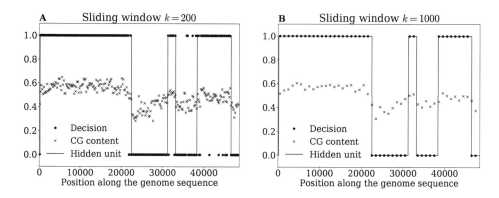

**Fig. 8.4** Best value of the hidden variable along the $\lambda$-phage sequence (full line), together with the percentage of $C$ and $G$ content averaged over a window of $k$ sites (crosses) and the Bayesian decision result (full circles) on the same window (**A.** $k = 200$, **B.** $k = 1000$).

# References

[1] Höhle, Michael and Held, Leonhard (2006). Bayesian estimation of the size of a population. *Discussion Paper, No. 499, Ludwig-Maximilians-Universität München, Sonderforschungsbereich 386 - Statistische Analyse diskreter Strukturen, München*.

[2] Berg, Howard C (1993). *Random walks in biology*. Princeton University Press.

[3] Ruthardt, Nadia, Lamb, Don C, and Bräuchle, Christoph (2011). Single-particle tracking as a quantitative microscopy-based approach to unravel cell entry mechanisms of viruses and pharmaceutical nanoparticles. *Molecular therapy*, **19**(7), 1199–1211.

[4] Robson, Alex, Burrage, Kevin, and Leake, Mark C (2013). Inferring diffusion in single live cells at the single-molecule level. *Phil. Trans. R. Soc. B*, **368**(1611), 20120029.

[5] Brune, Douglas and Kim, Sangtae (1993). Predicting protein diffusion coefficients. *Proceedings of the National Academy of Sciences*, **90**(9), 3835–3839.

[6] MacKay, David JC (2003). *Information theory, inference and learning algorithms*. Cambridge University Press.

[7] Jaynes, Edwin T (1957). Information theory and statistical mechanics. *Physical Review*, **106**(4), 620.

[8] van Steveninck, Rob R de Ruyter, Lewen, Geoffrey D, Strong, Steven P, Koberle, Roland, and Bialek, William (1997). Reproducibility and variability in neural spike trains. *Science*, **275**(5307), 1805–1808.

[9] Koch, Kristin, McLean, Judith, Berry, Michael, Sterling, Peter, Balasubramanian, Vijay, and Freed, Michael A (2004). Efficiency of information transmission by retinal ganglion cells. *Current Biology*, **14**(17), 1523–1530.

[10] Butts, Daniel A, Weng, Chong, Jin, Jianzhong, Yeh, Chun-I, Lesica, Nicholas A, Alonso, Jose-Manuel, and Stanley, Garrett B (2007). Temporal precision in the neural code and the timescales of natural vision. *Nature*, **449**(7158), 92.

[11] Schneidman, Elad, Berry II, Michael J, Segev, Ronen, and Bialek, William (2006). Weak pairwise correlations imply strongly correlated network states in a neural population. *Nature*, **440**(7087), 1007.

[12] Wishart, John (1928). The generalised product moment distribution in samples from a normal multivariate population. *Biometrika*, **20**(1/2), 32–52.

[13] Mehta, Madan Lal (2004). *Random matrices*. Elsevier.

[14] Bun, Joël, Bouchaud, Jean-Philippe, and Potters, Marc (2017). Cleaning large correlation matrices: tools from random matrix theory. *Physics Reports*, **666**, 1–109.

[15] Livan, Giacomo, Novaes, Marcel, and Vivo, Pierpaolo (2018). *Introduction to random matrices: theory and practice*. Springer.

[16] Marčenko, Vladimir A and Pastur, Leonid Andreevich (1967). Distribution of eigenvalues for some sets of random matrices. *Mathematics of the USSR-Sbornik*, **1**(4), 457.

[17] Watkin, T.L.H. and Nadal, J-P. (1993). Optimal unsupervised learning. *J. Phys. A: Math. and Gen*, **27**, 1899.

[18] Reimann, P, Van den Broeck, C, and Bex, GJ (1996). A Gaussian scenario for unsupervised learning. *Journal of Physics A* (29), 3521–3535.

[19] Peyrache, Adrien, Khamassi, Mehdi, Benchenane, Karim, Wiener, Sidney I, and Battaglia, Francesco P (2009). Replay of rule-learning related neural patterns in the prefrontal cortex during sleep. *Nature neuroscience*, **12**(7), 919–926.

[20] Peyrache, Adrien, Benchenane, Karim, Khamassi, Mehdi, Wiener, Sidney I, and Battaglia, Francesco P (2010). Principal component analysis of ensemble recordings reveals cell assemblies at high temporal resolution. *Journal of computational neuroscience*, **29**(1-2), 309–325.

[21] Montanari, A and Richard, E (2016). Non-negative principal component analysis: Message passing algorithms and sharp asymptotics. *IEEE Transactions on Information Theory* (62), 1458–1484.

[22] Monasson, Rémi (2017). Inference of principal components of noisy correlation matrices with prior information. *2016 50th Asilomar Conference on Signals, Systems and Computers*.

[23] Bernard, Elsa, Jacob, Laurent, Mairal, Julien, and Vert, Jean-Philippe (2014). Efficient RNA isoform identification and quantification from RNA-Seq data with network flows. *Bioinformatics*, **30**(17), 2447–2455.

[24] Friedman, Jerome, Hastie, Trevor, and Tibshirani, Robert (2008). Sparse inverse covariance estimation with the graphical lasso. *Biostatistics*, **9**(3), 432–441.

[25] Ackley, David H, Hinton, Geoffrey E, and Sejnowski, Terrence J (1985). A learning algorithm for Boltzmann machines. *Cognitive science*, **9**(1), 147–169.

[26] Haldane, Allan and Levy, Ronald M (2021). Mi3-GPU: MCMC-based inverse ising inference on GPUs for protein covariation analysis. *Computer Physics Communications*, **260**, 107312.

[27] Hinton, Geoffrey E (2002). Training products of experts by minimizing contrastive divergence. *Neural computation*, **14**(8), 1771–1800.

[28] Decelle, Aurélien, Furtlehner, Cyril, and Seoane, Beatriz (2021). Equilibrium and non-equilibrium regimes in the learning of restricted Boltzmann machines. *Advances in Neural Information Processing Systems*, **34**, 5345–5359.

[29] Mozeika, Alexander, Dikmen, Onur, and Piili, Joonas (2014). Consistent inference of a general model using the pseudolikelihood method. *Physical Review E*, **90**, 010101.

[30] Meinshausen, Nicolai and Bühlmann, Peter (2006). High-dimensional graphs and variable selection with the lasso. *The Annals of Statistics*, **34**(3), 1436–1462.

[31] Bateman, Alex, Coin, Lachlan, Durbin, Richard, Finn, Robert D, Hollich, Volker, Griffiths-Jones, Sam, Khanna, Ajay, Marshall, Mhairi, Moxon, Simon, Sonnhammer, Erik LL, Studholme, David J., Yeats, Corin, and Eddy, Sean R. (2004). The Pfam protein families database. *Nucleic acids research*, **32**, D138–D141.

[32] Cristianini, Nello and Hahn, Matthew W (2006). *Introduction to computational*

*genomics: a case studies approach.* Cambridge University Press.

[33] Durbin, Richard, Eddy, Sean R, Krogh, Anders, and Mitchison, Graeme (1998). *Biological sequence analysis: probabilistic models of proteins and nucleic acids.* Cambridge University Press.

[34] Cocco, Simona, Feinauer, Christoph, Figliuzzi, Matteo, Monasson, Rémi, and Weigt, Martin (2018). Inverse statistical physics of protein sequences: a key issues review. *Reports on Progress in Physics*, **81**(3), 032601.

[35] Morcos, Faruck, Pagnani, Andrea, Lunt, Bryan, Bertolino, Arianna, Marks, Debora S, Sander, Chris, Zecchina, Riccardo, Onuchic, José N, Hwa, Terence, and Weigt, Martin (2011). Direct-coupling analysis of residue coevolution captures native contacts across many protein families. *Proceedings of the National Academy of Sciences*, **108**(49), E1293–E1301.

[36] Jumper, John et al. (2021). Highly accurate protein structure prediction with AlphaFold. *Nature*, **596**(7873), 583–589.

[37] Du, Zongyang, Su, Hong, Wang, Wenkai, Ye, Lisha, Wei, Hong, Peng, Zhenling, Anishchenko, Ivan, Baker, David, and Yang, Jianyi (2021). The trRosetta server for fast and accurate protein structure prediction. *Nature protocols*, **16**(12), 5634–5651.

[38] Barton, John P, Cocco, Simona, De Leonardis, E, and Monasson, Rémi (2014). Large pseudocounts and $l_2$-norm penalties are necessary for the mean-field inference of Ising and Potts models. *Physical Review E*, **90**(1), 012132.

[39] Friedman, Jerome, Hastie, Trevor, and Tibshirani, Robert (2001). *The elements of statistical learning.* Springer series in statistics New York.

[40] Dunn, Stanley D, Wahl, Lindi M, and Gloor, Gregory B (2008). Mutual information without the influence of phylogeny or entropy dramatically improves residue contact prediction. *Bioinformatics*, **24**(3), 333–340.

[41] Harsh, Moshir, Tubiana, Jérôme, Cocco, Simona, and Monasson, Rémi (2020). 'Place-cell' emergence and learning of invariant data with restricted Boltzmann machines: breaking and dynamical restoration of continuous symmetries in the weight space. *Journal of Physics A: Mathematical and Theoretical*, **53**(17), 174002.

[42] Poggio, Tomaso and Anselmi, F. (2016). *Visual Cortex and Deep Networks: Learning Invariant Representations.* The MIT Press, Cambridge, MA, USA.

[43] Fanthomme, Arnaud, Rizzato, F, Cocco, Simona, and Monasson, Rémi (2022). Optimal regularizations for data generation with probabilistic graphical models. *Journal of Statistical Mechanics: Theory and Experiment*, **2022**(5), 053502.

[44] Oord, Aäron van den, Kalchbrenner, Nal, Vinyals, Oriol, Espeholt, Lasse, Graves, Alex, and Kavukcuoglu, Koray (2016). Conditional image generation with pixelCNN decoders. In *Proceedings of the 30th International Conference on Neural Information Processing Systems*, pp. 4797–4805.

[45] Oord, Aaron van den, Dieleman, Sander, Zen, Heiga, Simonyan, Karen, Vinyals, Oriol, Graves, Alex, Kalchbrenner, Nal, Senior, Andrew, and Kavukcuoglu, Koray (2016). Wavenet: A generative model for raw audio. *arXiv:1609.03499*.

[46] Wu, Dian, Wang, Lei, and Zhang, Pan (2019). Solving statistical mechanics using variational autoregressive networks. *Physical Review Letters*, **122**(8), 080602.

[47] Sharir, Or, Levine, Yoav, Wies, Noam, Carleo, Giuseppe, and Shashua, Amnon (2020). Deep autoregressive models for the efficient variational simulation of many-body quantum systems. *Physical Review Letters*, **124**(2), 020503.

[48] Doersch, Carl (2016). Tutorial on variational autoencoders. *arXiv:1606.05908*.

[49] Kingma, Diederik P, Welling, Max et al. (2019). An introduction to variational autoencoders. *Foundations and Trends in Machine Learning*, **12**(4), 307–392.

[50] Dayan, P., Hinton, G.E., Neal, R.M., and Zemel, R.S. (1995). The Helmholtz machine. *Neural computation*, **7**, 889–904.

[51] Goodfellow, Ian J, Pouget-Abadie, Jean, Mirza, Mehdi, Xu, Bing, Warde-Farley, David, Ozair, Sherjil, Courville, Aaron, and Bengio, Yoshua (2014). Generative adversarial networks. *arXiv:1406.2661*.

[52] Ekeberg, Magnus, Hartonen, Tuomo, and Aurell, Erik (2014). Fast pseudolikelihood maximization for direct-coupling analysis of protein structure from many homologous amino-acid sequences. *Journal of Computational Physics*, **276**, 341–356.

[53] Barton, John P, De Leonardis, Eleonora, Coucke, Alice, and Cocco, Simona (2016). ACE: adaptive cluster expansion for maximum entropy graphical model inference. *Bioinformatics*, **32**(20), 3089–3097.

[54] Figliuzzi, Matteo, Barrat-Charlaix, Pierre, and Weigt, Martin (2018). How pairwise coevolutionary models capture the collective residue variability in proteins? *Molecular biology and evolution*, **35**(4), 1018–1027.

[55] McGee, Francisco, Hauri, Sandro, Novinger, Quentin, Vucetic, Slobodan, Levy, Ronald M, Carnevale, Vincenzo, and Haldane, Allan (2021). The generative capacity of probabilistic protein sequence models. *Nature communications*, **12**(1), 1–14.

[56] Figliuzzi, Matteo, Jacquier, Hervé, Schug, Alexander, Tenaillon, Oliver, and Weigt, Martin (2016). Coevolutionary landscape inference and the context-dependence of mutations in beta-lactamase TEM-1. *Molecular biology and evolution*, **33**(1), 268–280.

[57] Hopf, Thomas A., Ingraham, John B., Poelwijk, Frank J., Schärfe, Charlotta P.I., Springer, Michael, Sander, Chris, and Marks, Debora S. (2017). Mutation effects predicted from sequence co-variation. *Nature biotechnology*, **35**, 128–135.

[58] Russ, William P, Figliuzzi, Matteo, Stocker, Christian, Barrat-Charlaix, Pierre, Socolich, Michael, Kast, Peter, Hilvert, Donald, Monasson, Remi, Cocco, Simona, Weigt, Martin et al. (2020). An evolution-based model for designing chorismate mutase enzymes. *Science*, **369**(6502), 440–445.

[59] Tubiana, Jérôme, Cocco, Simona, and Monasson, Rémi (2019). Learning protein constitutive motifs from sequence data. *Elife*, **8**, e39397.

[60] Bravi, Barbara, Tubiana, Jérôme, Cocco, Simona, Monasson, Rémi, Mora, Thierry, and Walczak, Aleksandra M (2021). RBM-MHC: A semi-supervised machine-learning method for sample-specific prediction of antigen presentation by HLA-I alleles. *Cell Systems*, **12**(2), 195–202.

[61] Riesselman, Adam J, Ingraham, John B, and Marks, Debora S (2018). Deep generative models of genetic variation capture the effects of mutations. *Nature methods*, **15**(10), 816–822.

[62] Shin, Jung-Eun, Riesselman, Adam J, Kollasch, Aaron W, McMahon, Conor, Simon, Elana, Sander, Chris, Manglik, Aashish, Kruse, Andrew C, and Marks, Debora S (2021). Protein design and variant prediction using autoregressive generative models. *Nature communications*, **12**(1), 1–11.

[63] Trinquier, Jeanne, Uguzzoni, Guido, Pagnani, Andrea, Zamponi, Francesco, and Weigt, Martin (2021). Efficient generative modeling of protein sequences using simple autoregressive models. *Nature Communications*, **12**, 5800.

[64] Sanger, T.D. (1989). Optimal unsupervised learning in a single-layer linear feedforward neural network. *Neural Networks*, **2**, 459–473.

[65] Cover, Thomas M (1965). Geometrical and statistical properties of systems of linear inequalities with applications in pattern recognition. *IEEE transactions on electronic computers* (3), 326–334.

[66] Gardner, Elizabeth (1988). The space of interactions in neural network models. *Journal of physics A: Mathematical and general*, **21**(1), 257.

[67] Altarelli, Fabrizio, Monasson, Rémi, Semerjian, Guilhem, and Zamponi, Francesco (2009). Connections to statistical physics. *Handbook of Satisfiability*, 569–611.

[68] Broeck, C. Van Den and Engel, A. (2001). *Statistical mechanics of learning.* Cambridge University Press.

[69] Mitchison, GJ and Durbin, RM (1989). Bounds on the learning capacity of some multi-layer networks. *Biological Cybernetics*, **60**(5), 345–365.

[70] Stiffler, Michael A, Chen, Jiunn R, Grantcharova, Viara P, Lei, Ying, Fuchs, Daniel, Allen, John E, Zaslavskaia, Lioudmila A, and MacBeath, Gavin (2007). PDZ domain binding selectivity is optimized across the mouse proteome. *Science*, **317**(5836), 364–369.

[71] Bellman, Richard E and Dreyfus, Stuart E (2015). *Applied dynamic programming.* Volume 2050. Princeton University Press.

[72] Viterbi, Andrew (1967). Error bounds for convolutional codes and an asymptotically optimum decoding algorithm. *IEEE transactions on Information Theory*, **13**(2), 260–269.

[73] Baum, Leonard E, Petrie, Ted, Soules, George, and Weiss, Norman (1970). A maximization technique occurring in the statistical analysis of probabilistic functions of markov chains. *The annals of mathematical statistics*, **41**(1), 164–171.

# Index